ROUTLEDGE LIBRARY EDITIONS: NUCLEAR SECURITY

Volume 29

NUCLEAR-FREE ZONES

NUCLEAR-FREE ZONES

Edited by
DAVID PITT AND GORDON THOMPSON

LONDON AND NEW YORK

First published in 1987 by Croom Helm

This edition first published in 2021
by Routledge
2 Park Square, Milton Park, Abingdon, Oxon OX14 4RN

and by Routledge
52 Vanderbilt Avenue, New York, NY 10017

Routledge is an imprint of the Taylor & Francis Group, an informa business

© 1987 David Pitt and Gordon Thompson

All rights reserved. No part of this book may be reprinted or reproduced or utilised in any form or by any electronic, mechanical, or other means, now known or hereafter invented, including photocopying and recording, or in any information storage or retrieval system, without permission in writing from the publishers.

Trademark notice: Product or corporate names may be trademarks or registered trademarks, and are used only for identification and explanation without intent to infringe.

British Library Cataloguing in Publication Data
A catalogue record for this book is available from the British Library

ISBN: 978-0-367-50682-7 (Set)
ISBN: 978-1-00-309763-1 (Set) (ebk)
ISBN: 978-0-367-53031-0 (Volume 29) (hbk)
ISBN: 978-1-00-308017-6 (Volume 29) (ebk)

Publisher's Note
The publisher has gone to great lengths to ensure the quality of this reprint but points out that some imperfections in the original copies may be apparent.

Disclaimer
The publisher has made every effort to trace copyright holders and would welcome correspondence from those they have been unable to trace.

NEW FOREWORD 2020

This book, edited by David Pitt and me, was first published in 1987. Sadly, David died in 2016. I know that he would have applauded the republication of this book. Continuing demand for the book indicates ongoing interest in strengthening the existing nuclear-weapon-free zones, and in creating new zones. That interest gives me renewed hope for humanity's shared future.

Since 1987, there have been many events around the world that relate directly or indirectly to NWFZs. The number and complexity of these events are too great for them to be summarized here. Fortunately, however, we now have a tool that allows anyone with Internet access to quickly obtain and share knowledge related to NWFZs. That tool, the World Wide Web, was invented in 1989. New opportunities for constructive use of the Web are emerging. Those opportunities could include research, education, campaigning, and diplomacy that lead to enhanced roles for NWFZs.

A major reason for celebration about NWFZs is their increased number. In 1987 there were two zones in inhabited areas: Latin America (Treaty of Tlatelolco, 1967); and the South Pacific (Treaty of Rarotonga, 1985). Now, there are three more: Southeast Asia (Treaty of Bangkok, 1995); Africa (Treaty of Pelindaba, 1996); and Central Asia (Treaty of Semipalatinsk, 2006). I hope and expect that readers of this book will identify ways to expand the list.

At the time of writing, the world is gripped by the COVID-19 pandemic that began in December 2019. I hope that this pandemic is speedily resolved. I further hope that this pandemic enhances awareness of humanity's shared vulnerability and potential for shared purpose, leading to outcomes that include reduction and elimination of nuclear arsenals.

Gordon Thompson
Institute for Resource and Security Studies
Cambridge, Massachusetts

March 2020

NUCLEAR-FREE ZONES

Edited by David Pitt and Gordon Thompson

CROOM HELM
London • New York • Sydney

© 1987 David Pitt and Gordon Thompson
Croom Helm Ltd, Provident House,
Burrell Row, Beckenham, Kent BR3 1AT
Croom Helm Australia, 44-50 Waterloo Road,
North Ryde, 2113, New South Wales

Published in the USA by
Croom Helm
in association with Methuen, Inc.
29 West 35th Street
New York, NY 10001

British Library Cataloguing in Publication Data

Nuclear-free zones.
  1. Nuclear-weapon-free zones
  I. Pitt, David II. Thompson, Gordon
  363.3'5   JX1974.735
  ISBN 0-7099-4076-9

Library of Congress Cataloging-in-Publication Data

ISBN 0-7099-4076-9

Printed and bound in Great Britain
by Billing & Sons Limited, Worcester.

CONTENTS

Preface - Alexandre Berenstein

1. Nuclear-Free Zones: An Idea Whose Time Has Come
   David Pitt — 1
2. Antarctica as a Nuclear-Free Zone
   Ramesh Thakur and Hyam Gold — 7
3. The Treaty for the Prohibition of Nuclear Weapons in Latin America
   Alfonso Garcia Robles — 9
4. The Treaty of Rarotonga: The South Pacific Nuclear-Free Zone
   Ramesh Thakur — 23
5. Regional Arms Control in the South Pacific
   Greg Fry — 46
6. For a Nuclear-Free Europe
   Ken Coates — 67
7. The Quest for a Balkan Nuclear-Weapons-Free Zone
   Peri Pamir — 94
8. A Nuclear-Weapon-Free Zone in Africa?
   William Epstein — 110
9. Nuclear-Free Zones: Problems and Prospects
   Ken Coates — 128

Conclusion
   Gordon Thompson — 138

About the Editors and Contributors — 140

About the Sponsoring Organisations — 141

Index — 143

This book is part of an occasional series on issues relating to nuclear arms control, sponsored by

GENEVA INTERNATIONAL PEACE RESEARCH INSTITUTE
Geneva, Switzerland

and

INSTITUTE FOR RESOURCE AND SECURITY STUDIES
Cambridge, Massachusetts, USA

ACKNOWLEDGEMENTS

The pieces by Ken Coates originally appeared in The Most Dangerous Decade (Bertrand Russell Foundation, 1984) and William Epstein's essay was originally a Stanley Foundation Occasional Paper (1977). We are grateful to the authors for permission to produce them here.

PREFACE

This book is one of the products of GIPRI's research programme to explore innovative approaches in the pursuit of peace. We are pleased to join in sponsoring this book with the Institute for Resource and Security Studies.
   The nuclear-free zone idea is not new but it is currently being regarded as a very hopeful mechanism for the future. Our book contains a selection of papers which explore the theory and practice of nuclear-free zones.
   The nuclear-free zone movement is growing rapidly as an alternative approach in the search for peace, and in some countries has support from the majority of people. This book is intended to document this growth of a movement, as well as providing students of peace with a reader that will help them in their studies.
   We invite those with relevant knowledge and experience to send us their thoughts so that future volumes can expand on these themes.

Professor Alexandre Berenstein
President
Geneva International Peace Research Institute

Chapter One

NUCLEAR-FREE ZONES : AN IDEA WHOSE TIME HAS COME

David Pitt

In 1985 there was a story going round the Press Bar in the Palais des Nations in Geneva just after the sinking of the Greenpeace ship, the RAINBOW WARRIOR in Auckland. At a diplomatic cocktail party the New Zealanders were lamenting to their French colleagues about the incident. 'At least' said the French diplomat 'everybody knows now where New Zealand is'. Nuclear-free zones (NFZ) are probably in the same category and partly for the same reason. Nuclear-free zones have become not only a political issue but also part of the public consciousness to the extent that people put NFZ stickers on their houses, cars and even their toilets. But what is a nuclear-free zone? Some people want no contact with any form of nuclear energy. The usage in international relations as adopted by the United Nations is not nearly so comprehensive. Nuclear-free zones are areas of the world where most of the countries of a given region want to be free of nuclear weapons, whilst not interfering with existing treaties or international legal freedoms and with the support of existing nuclear powers. But what a nuclear weapon is, has not been clearly defined nor are peaceful uses (likewise not defined) necessarily prohibited at all. Some countries allow the passage of nuclear weapons and others go further than weapons in prohibiting, for example, the dumping of radioactive wastes. This book is concerned with nuclear-weapons-free zones, in this limited sense.
   The idea of nuclear-weapon-free zones is as old as nuclear weapons. The horrors of Hiroshima and Nagasaki produced an instant reaction. The very first resolution of the United Nations made recommendations for the international control of atomic energy and its use for peaceful purposes. There has long been a particular fear amongst those nations which were not part of the nuclear club. In the 1950s when the idea was widely discussed an initial impetus came from the Central European nations. Adam Rapacki from Poland, a country often overrun by bellicose neighbours, produced a plan for an 'atomic free' Central Europe involving the removal

NUCLEAR-FREE ZONES: AN IDEA WHOSE TIME HAS COME

of nuclear weapons from East and West Germany, Poland and Czechoslovakia.
There were early moves too in the Baltic, where there was a strong neutral current. Finland proclaimed itself nuclear-free as early as 1947. In 1961, at the United Nations, Sweden tabled the Unden Plan calling for, inter alia, the establishment of nuclear-free zones. Elsewhere there were similar suggestions in the 1950s and 60s - Balkans, Mediterranean, China, Indian Ocean, Ireland, Antarctica and Latin America. In the last 2 areas treaties were produced (1959, 1967). The Latin American Treaty of Tlatelolco (1967) was a particularly significant document. The contracting parties undertook to prohibit testing, manufacture or storage of any kind of nuclear weapon or devices for launching them and direct or indirect acquisition. By 1985 there are still many problems and deficiencies - Cuba has not signed (and will not until the USA returns Guatanamo) nor some Carribean Countries, nor Guyana (because of a border dispute with Venezuela). Argentina has not ratified the treaty and Brazil and Chile, though ratifying have not fulfilled the conditions for entry into force. Nonetheless, Tlatelolco was the first, and until the Rarotonga Treaty in 1985, the only treaty covering a populated area.

However it was initially not only smaller countries but also superpowers who sought nuclear-free zones. The nuclear club (initially the USA, USSR, and UK, later joined by France and China) was and remains most anxious to prevent 'horizontal' proliferation to other powers. From 1962 there were long running discussions in the 18-nation (later enlarged) Committee on Disarmament, under the permanent co-chairmanship of the USA and USSR, from which emerged a Non-Proliferation Treaty (NPT) (1968). By 1985, one hundred and thirty countries had become party to that treaty which encouraged (in Article VII) groups of states to conclude treaties which would assure the total absence of nuclear weapons from their region.

The NPT has been a 'curate's egg' - good in parts. It has not prevented so-called 'vertical proliferation', the accumulation of more weapons by members of the nuclear club. The number of such weapons in the American and Soviet arsenals has risen from approximately 6,000 in 1970 to 20,000 in 1985. China and France have not signed. Other powers almost certainly have nuclear devices (India, Pakistan, South Africa, Israel) and a number of others are on, and can easily pass over, the threshold (Argentina, Brazil, Iraq, South Korea, Taiwan). In short, perhaps three quarters of the world's population live in nuclear or near nuclear states. Adding to the process of horizontal proliferation have been the increasing deployment of superpower weaponry in friendly states and the transfer of nuclear technology generally. The civil use of nuclear energy is considered a relatively easy

NUCLEAR-FREE ZONES: AN IDEA WHOSE TIME HAS COME

route to an eventual acquisition. Groups of schoolboys have even been designing devices. In a 1985 briefing document on the NPT, Greenpeace has therefore concluded that by the end of the 1980s 'all but the poorest third world countries will have some sort of nuclear plant and technology'.

Despite, probably because of, the troubles of the NPT and NFZs the NFZ movement has taken on a fresh vigour and vitality in the 1980s. There has been a great upsurge of popular antipathy to things nuclear. In Australia and New Zealand local authorities and communities began to declare themselves nuclear-free. According to the Nuclear Free Zone Registry there were by mid-1984 over 1,600 cities in the world which were NFZs, many in Britain and over 80 in the USA, totalling over 10 million people in that country alone. Of course the degree to which these cities were, or could be 'free' varied and some were only able to restrict transport of materials etc., or simply make the protest.

Nonetheless the mushrooming of such NFZs was a sign of a strong public opinion. In one country (New Zealand) nearly two thirds of the population lived in areas which had declared themselves NFZs. These people in fact voted in a Labour Government in 1984 on a nuclear-free platform. In 1985 a new treaty in the South Pacific emerged, initiated not only by Australia and New Zealand but by many small island nations who were regularly confronted by French nuclear tests on the coral island of Mururoa. Trade unions in the Pacific banded together to prohibit nuclear ships and cargoes from ports.

The nuclear club reaction to the recent emergence of NFZs has been mixed and in some cases hostile, as in the French bombing of the Rainbow Warrior. The USA has reacted strongly to the prohibition of its nuclear ships in New Zealand and the Pacific and has even purportedly 'invaded' a former Pacific territory (Palau) which has established a nuclear-free constitution. There is a fear of a domino effect among US allies in Asia and the Pacific. At the NPT Review Conference in 1985 the USA and the UK were most reluctant to see any curtailing of their nuclear prerogatives, thus allowing the USSR, which has espoused the NFZ idea, to curry favour with the non-aligned bloc. More generally perhaps, there is a fear and suspicion amongst power elites that ideas of NFZs supported by 'green' parties and movements, especially amongst young people and women, represent a political threat. This book represents a first attempt to provide some kind of analysis of this important area by looking at the international aspect - first the treaties that have been concluded involving terrestrial areas (Antarctica, Tlatelolco, Rarotonga). We have not included a discussion of the sea-bed, outer space or subnational nuclear-free zones, though we hope to treat these in later volumes. We have however included a number of essays that discuss regions where nuclear-free zones have been suggested (Europe,

NUCLEAR-FREE ZONES: AN IDEA WHOSE TIME HAS COME

Balkans and Africa) and where treaties may emerge in the future, especially after the Rarotonga publicity. The NPT discussions in Geneva in 1985 after all showed that the group of non-aligned nations are closing ranks on this as well as economic issues through the Group of 77. This Southern Coalition is also finding an ally in Northern nations outside the superpower complex. There is an increasing momentum behind discussions in both Western (and especially Northern) Europe and the Balkans. The superpowers themselves may be, albeit grudgingly, recognising a global situation where there is no one power bloc or superpower domination.

The book then, we hope, is part of a wave of the future. Significantly, most of our contributors come from small countries, though this may be more by accident than design, since in the Soviet Union at least, nuclear-free zones have been studied for a long time. More important perhaps, our contributors come from a wide range of disciplines and from academic, activist and political backgrounds. There is a little doubt that we are witnessing a reshuffling of academic boundaries as well as a crossing of the line between the pure and the applied. A new interdisciplinary, action-oriented kind of peace research is emerging, based on rational analysis, legal argument and solid scientific study, far removed from ideological rhetoric. We hope this book will be of value to the growing numbers of peace studies courses that are being mounted in not only the universities but also a wide range of schools and educational institutions. We hope our audience will be wider still. Peace, once the most important part of the United Nations work until overshadowed by the emphasis on development, may again become central, if only because the success of peace and development are inextricably intertwined. In the labyrinth of arms negotiations, clearly written analyses are desperately needed by those directly involved, as well as by the journalists who relay the deliberations and results on to the wider public. We hope this book will be useful to all these people.

We have adopted a chronological approach. We begin with Thakur and Gold's brief analysis of the first nuclear-free zone treaty in Antarctica. In many ways this treaty was exceptional, applying to a largely uninhabited area where national claims were (in one area) either non-existent or at least initially without significant exploitation. In this sense the Antarctic Treaty belongs with later negotiations over the sea-bed and outer space. But the idea begun in Antarctica generated a momentum from the late sixties onward, noticeably in those areas that were contiguous to it, that is Latin America and the South Pacific.

We continue therefore with Garcia Robles' discussion and defence of the Tlatelolco Treaty. Robles is important amongst our authors in that he was an active negotiator, if not the father of a treaty. Whatever the critics have said about

NUCLEAR-FREE ZONES: AN IDEA WHOSE TIME HAS COME

Tlatelolco it was for nearly 20 years a unique institution in itself. Not until the mid-1980s do we find a new treaty - different and more comprehensive even if it also had an exotic, albeit more easily pronounced name - Rarotonga. We present two complementary views of the Rarotonga Treaty from New Zealand (Thakur) and Australia (Fry). Both are concerned to show what the Rarotonga Treaty can and cannot do in a situation of superpower indifference or hostility.

The second half of our book is concerned with 'treaties yet to be', with proposals and movements, particularly in three areas, Western Europe, the Balkans and Africa. Of these, the earliest moves were in Europe. Coates' paper is a plea for a nuclear-free zone in Europe, continuing the important ideas of statesmen like Rapacki, and more recently, Palme. We then continue with Pamir and the Balkans where there is also a long history of attempts to move towards a nuclear-free zone, dating back to at least the Roumanian initiative of 1957. Then follows an essay from Epstein on Africa, a continent which has suffered in recent years both severe development crises and many of the 300 or more conventional wars which have plagued mankind since World War II. Proposals for a nuclear-free zone in Africa were not much behind Europe, following the French tests in the Sahara in 1960. We conclude the essays from our contributors with Coates looking at the general problems and prospects for nuclear-free zones.

Our conclusions can only be interim. Whilst the number of nuclear weapons continues daily to escalate and horizontal proliferation seems impossible to prevent, the idea of the nuclear-free zone has a greater urgency and potential application (e.g. conservation zones). But the idea has itself no teeth and indeed there are those who say, that the treaties that exist are not enforceable or respected or have never been put to the test. The history of international relations is regrettably littered with treaties whose provisions have been broken, sometimes before the ink is dry on them. Will this be the fate of the nuclear-free zones treaties or will the growing weight of public opinion press on governments who seem either not to want arms reduction, or to be unable to achieve it? Nuclear-free zones may in the ultimate analysis be successful precisely because they are a concrete and tangible way for people to show their aspirations for peace. There can no longer be the 'lambs and dragons' situation that David Low (originally from New Zealand, a country where sheep outnumbered people twenty times over) portrayed in a famous cartoon now displayed in a glass case in the League of Nations Museum in Geneva. The cartoon, drawn after a disarmament conference in the 1930s, showed a stage on which a nasty assortment of animals, dragons and even a crocodile, weeping crocodile tears, were talking to an audience of peaceful looking sheep. 'We regret to say' said the chief

NUCLEAR-FREE ZONES: AN IDEA WHOSE TIME HAS COME

dragon 'that we have been unable to control your warlike passions'. Through the nuclear-free passion the sheep may yet strike back.

Postscript

Since this book was written the quorum of 8 local nations and two superpowers, the People's Republic of China and the USSR, have signed the Rarotonga Treaty; the last with some reservations. Moreover, Mr Gorbatchev, as part of sweeping reforms in Soviet foreign, as well as domestic, policy has proposed a nuclear-weapon-free world by the year 2000. Even if the chances of achieving this utopia seem remote, given the refusal of the USA and France to acknowledge nuclear-weapon-free zones and indeed their active policies to increase nuclear weaponary, we may be entering a new phase of international relations where the small nuclear-free champions are joined more actively by some of the big boys.

Chapter Two

ANTARCTICA AS A NUCLEAR-FREE ZONE

Ramesh Thakur and Hyam Gold

The Antarctic Treaty is of great historical significance in the movement towards establishing and extending nuclear-free zones: it was the first such zone to be created in the world. After the adoption of the Antarctic Treaty at a conference in Washington on 1 December 1959, commentators hailed it as a model of international cooperation. Indeed some speculated on the relevance of the model for talks on disarmament, demilitarising other regions of the world, and even the possibility of a regime for outer space.

Signed by all twelve participating governments at the conclusion of the Washington Conference, the Antarctic Treaty has been in force since 23 June 1961. There are now 16 Antarctic Treaty consultative parties. Argentina, Australia, Belgium, Britain, Chile, France, Japan, New Zealand, Norway, South Africa, USA and the USSR are the twelve original signatories. Poland became a consultative party in 1977, West Germany in 1981, and Brazil and India in 1983. In addition there are 12 'acceding' states which have signed the treaty but are yet to undertake the substantial scientific activity necessary for admission to consultative status; they are: Bulgaria, China, Czechoslovakia, Denmark, East Germany, Italy, the Netherlands, Papua New Guinea, Peru, Rumania, Spain and Uruguay.

The many substantive and continuing achievements of the Antarctic Treaty owe much to the fact that the regime consists of both static and dynamic elements. The treaty seeks to demilitarise and denuclearise Antarctica and provide freedom of scientific research while holding in perpetual abeyance both claims to sovereignty and disputes about such claims. The static elements of demilitarisation and denuclearisation are complemented by the dynamic element of cooperative arrangements in research, exploration and environmental protection. The instrument which gives practical effect to the system is the periodic consultative meeting held under Article IX of the treaty. Membership in the Antarctic Club is open; any UN member can become a party to the treaty. Moreover, any

ANTARCTICA AS A NUCLEAR-FREE ZONE

country engaged in substantive scientific research in Antarctica can become a consultative party, and would find it in its interests to do so.

The basic rules of the Antarctic Treaty aim at demilitarising the continent and opening it for free scientific research. The demilitarisation and denuclearisation provisions are contained in Articles I and V. Participants at the Washington Conference had agreed on the need to reserve Antarctica exclusively for peaceful purposes, but differed on the wording of the clauses. The final phrasing of Article I emerged from a Soviet draft. After declaring that Antarctica shall be used for peaceful purposes only, Article I prohibits any measures of a military nature, e.g. military bases, fortifications and manoeuvres, and weapons testing. Military personnel and equipment may however be used for research or other peaceful purposes. Not all countries at the 1959 Conference were happy with this significant exception to the demilitarisation principle, but accepted it as a concession to practical reality. Article V prohibits any nuclear explosions and the disposal of radioactive waste materials in Antarctica, the latter provision being subject to the possibility of being overtaken by a future multilateral agreement. In 1959, it was the nuclear powers who favoured allowing peaceful nuclear explosions with appropriate safeguards. The argument failed to impress the non-nuclear parties, who remained firm in having the treaty impose a total ban on all nuclear explosions.

The Antarctic Treaty has had enviable successes in: demilitarising and denuclearising an entire continent, with an effective system of onsite inspection to back it up; freezing chauvinism; instituting a free flow of cooperative scientific research; and adopting what are by historical standards rather stringent measures for safeguarding the environment.

There is nevertheless a blot on the otherwise unblemished anti-nuclear record of the Antarctic Treaty. The United States operated a nuclear reactor at McMurdo Station in the Ross Dependency from 1969 until 1972, for the purpose of generating electricity. The reactor and radioactive material from McMurdo were shipped out between 1972 and 1979, at which time the reactor site at McMurdo was released for unrestricted use. The reactor and contaminated material were shipped to the USA for disposal in South Carolina and California. Thus Antarctica has been a totally nuclear free zone in the 1980s.

NOTES

This essay is adapted from the authors' 'The Antarctic Treaty Regime: Exclusive Preserve or Common Heritage?', Foreign Affairs Reports, Vol. XXXII, Nos. 11 and 12, November-December 1983, pp. 169-86.

Chapter Three

THE TREATY FOR THE PROHIBITION OF NUCLEAR WEAPONS IN LATIN AMERICA

Alfonso Garcia Robles

It seems advisable to point out from the outset that the Latin American nuclear-weapon-free zone had the privilege for many years of being the only one in existence which covered densely inhabited territories. Outside it there was only Antarctica, Outer Space and the Sea Bed where similar prohibitions are in force based on three different treaties concluded in 1959, 1967 and 1971, respectively.

The official title of the treaty which established the Latin American zone and defined its statute is 'Treaty for the prohibition of nuclear weapons in Latin America' but it is usually referred to as 'Treaty of Tlatelolco' employing the Aztec name of the district of the Mexican capital where the Ministry of Foreign Affairs of Mexico is located and where the treaty itself was opened for signature on 14 February 1967.

A study in depth of the subject which I have chosen would be obviously impossible in a single essay. My intention is, therefore, to provide only a synoptic view both of the genesis of the treaty and of the most salient features which the analysis of its provisions may reveal.

GENESIS OF THE TREATY OF TLATELOLCO

The first international document in the history of the events directly related to the genesis of the Treaty of Tlatelolco is the Joint Declaration of 29 April 1963 in which the presidents of Bolivia, Brazil, Chile, Ecuador and Mexico announced that their governments were willing to sign a Latin American multilateral agreement under which they would undertake not 'to manufacture, store, or test nuclear weapons or devices for launching nuclear weapons'.

Seven months later, the United Nations General Assembly, taking as a basis a draft resolution submitted by eleven Latin American countries (the five previously mentioned plus Costa Rica, El Salvador, Haiti, Honduras, Panama and Uruguay) approved, on 27 November 1963, resol-

TREATY: PROHIBITION OF NUCLEAR WEAPONS

ution 1911 (XVIII) in which inter alia it welcomed the initiative of the five presidents for the military denuclearisation of Latin America; expressed the hope that the States of the region would initiate studies 'concerning the measures that should be agreed upon with a view to achieving the aims' of the joint declaration, and requested the Secretary General of the United Nations to extend to the States of Latin America, at their request, 'such technical facilities as they may require in order to achieve the aims set forth in the present resolution'.

Almost one year elapsed between the adoption of this General Assembly resolution and the next step worth mentioning in a review of the antecedents of the treaty. This interval was not wasted however. The Mexican Government put it to good use with active diplomatic consultations which resulted in the convening of a Latin American conference named 'Preliminary Session on the Denuclearisation of Latin America' (known as REUPRAL, its Spanish acronym), which met in Mexico City from 23 to 27 November 1964, and which adopted a measure which was later to prove decisive for the success of the Latin American enterprise: the creation of an ad hoc organ, the 'Preparatory Commission for the denuclearisation of Latin America' (known also, as the previous one, by its Spanish acronym, which in this case was COPREDAL). The Preparatory Commission was specifically instructed, in the resolution by which it was established, 'to prepare a preliminary draft of a multilateral treaty for the denuclearisation of Latin America, and to this end, to conduct any prior studies and take any prior steps that it deems necessary'.

COPREDAL had its first session in Mexico City from 15 to 22 March 1965. In this session, the Commission adopted its rules of procedure and set up four subsidiary organs: a Coordination Committee and three working groups. Subsequently the Commission would create another subsidiary organ entitled 'Negotiating Committee'.

The Preparatory Commission held a total of four sessions. The last of them took place, just under two years after its creation, from 31 January to 14 February 1967. Contrary to what has generally happened with other disarmament treaties and conventions, the draft articles for the future treaty dealing with verification, inspection and control were the first to be completed at the second session of the Commission (23 August – 2 September 1965) when a full declaration of principles was also drafted to serve as a basis for the preamble of the draft treaty.

During its third session, COPREDAL received from its Coordinating Committee a working paper which contained the complete text of a preliminary draft treaty. This draft, together with other proposals submitted by member States, provided the basis for the deliberations of the session. The

TREATY: PROHIBITION OF NUCLEAR WEAPONS

result was the unanimous approval of a document entitled 'Proposals for the Preparation of the Treaty for the Denuclearisation of Latin America' which played as prominent a role in the history of the treaty as that of the Dumbarton Oaks proposals in the history of the United Nations. These 'Proposals' included all provisions which might prove necessary for the treaty as a whole, although in some cases COPREDAL, not having been able to find solutions satisfactory to all, had been obliged to present to the Governments two parallel alternatives.

Of those few pending questions which the Commission would be called upon to solve at its fourth session, the most important one was the entry into force of the treaty, probably the issue which had provoked the greatest discussion in its proceedings. Both because of this reason and because of the positive influence which the solution given to the problem involved may have in future similar cases, it is worthwhile to examine it in somewhat greater detail.

When the Preparatory Commission considered this subject in April 1966, two distinct viewpoints became apparent. According to the first, the treaty should come into force between States which would ratify it, on the date of deposit of their respective instruments of ratification, in keeping with the standard practice. The representative Latin American body which would be established by the treaty should begin to function as soon as eleven instruments of ratification were deposited, as this number constituted a majority of the twenty-one members of the Preparatory Commission. Those States supporting the alternative view, on the other hand, argued that the treaty, although signed and ratified by all Member States of the Preparatory Commission, should enter into force only upon completion of four requirements, essentially those defined in Article 28 of the treaty, which may be summarised as follows: the signature and ratification of the Treaty of Tlatelolco and its Additional Protocols I and II by all States to which they were opened and the conclusion of bilateral or multilateral agreements concerning the application of the Safeguards System of the International Atomic Energy Agency (IAEA) by each party to the treaty.

Consequently, this was one of the items on which, as I have already pointed out, COPREDAL was obliged to present in its proposals two parallel texts. These texts stated respectively the provisions that the treaty would contain, according to whether one accepted the first or second thesis. To solve the problem, the Coordinating Committee, in its report of 28 December 1966, suggested the adoption of a conciliatory formula, which could receive the approval of all Member States of the Commission without detriment to their respective positions on the alternative texts. It was this formula, with some modifications, which was finally to be adopted and incorporated into Article 28 of the treaty. In

TREATY: PROHIBITION OF NUCLEAR WEAPONS

keeping with it, the treaty would go into effect for all States that had ratified it upon completion of the four requirements specified in paragraph 1 of the article mentioned above. That notwithstanding, as the second paragraph of the article states:

> All signatory States shall have the imprescriptible right to waive, wholly or in part, the requirements laid down in the preceding paragraph. They may do so by means of a declaration which shall be annexed to their respective instrument of ratification and which may be formulated at the time of deposit of the instrument or subsequently. For those States which exercise this right, this Treaty shall enter into force upon deposit of the declaration, or as soon as those requirements have been met which have not been expressly waived.

The third paragraph of the same article stipulates, moreover:

> As soon as this Treaty has entered into force in accordance with the provisions of paragraph 2 for eleven States, the Depositary Government shall convene a preliminary meeting of those States in order that the Agency may be set up and commence its work.

As one can see, an eclectic system was adopted, which, while respecting the viewpoints of all signatory States, prevented nonetheless any particular State from precluding the enactment of the treaty for those which would voluntarily wish to accept the statute of military denuclearisation defined therein.

The Treaty of Tlatelolco has thus contributed effectively to dispel the myth that, for the establishment of a nuclear-weapon-free zone it would be an essential requirement that all States of the region concerned should become, from the very outset, parties to the treaty establishing the zone. The system adopted in the Latin American instrument proves that, although no State can obligate another to join such a zone, neither can one prevent others wishing to do it from adhering to a regime of total absence of nuclear weapons within their own territories.

Once the question of the entry into force of the treaty had been settled, at the fourth session of COPREDAL, in the manner just explained, the Preparatory Commission proceeded to settle also without major difficulties the few other pending problems. On 12 February 1967, the Treaty for the Prohibition of Nuclear Weapons in Latin America was unanimously approved and two days later, at the solemn closing ceremony of the Commission's proceedings, it was opened for signature and subscribed to the same day by the representatives of

TREATY: PROHIBITION OF NUCLEAR WEAPONS

fourteen of its twenty-one members. As of today, the signatory States have reached the number of twenty-six, of which twenty-three are already parties to the treaty.

Additional Protocol I which is open to the four States - United Kingdom, Netherlands, United States and France - which are internationally responsible for territories lying within the limits of the geographical zone established in the treaty, has been signed and ratified by the first three of those States. France has also signed the Protocol, but it has not yet ratified it.

As to additional Protocol II, the five nuclear-weapon States have already become parties to it in the following chronological order: United Kingdom, United States, France, China and USSR.

As provided in paragraph 3 of Article 28 previously quoted, as soon as the treaty entered into force for eleven States, the Depositary Government convened a 'preliminary meeting' of those States in order to set up the Agency for the Prohibition of Nuclear Weapons in Latin America, known by its Spanish acronym OPANAL. This preliminary meeting (REOPANAL) took place in late June 1969 and carried out successfully all the preparatory work necessary for the first session of the General Conference of OPANAL. The latter was inaugurated on 2 September 1969 in the presence of U Thant, the then Secretary-General of the United Nations, and the Director General of the IAEA, Sigvard Eklund. After seven working days, the General Conference gave its approval to a series of basic juridical and administrative documents which provided the foundations for the Latin American Agency created by the treaty. To this date, the General Conference has held eight regular sessions and two special sessions in accordance with the provisions of Article 9.

ANALYTICAL SUMMARY OF THE TREATY

Following the brief survey I have just made of the preparatory work leading to the conclusion of the Tlatelolco Treaty, I will attempt to give now a general idea of its contents and a brief analytical summary of some of its main provisions.

The treaty comprises a preamble, thirty-one articles, one transitional article and two additional protocols.

The preamble defines the fundamental aims pursued by the States which drafted the treaty by stating their conviction that:

> The military denuclearisation of Latin America - being understood to mean the undertaking entered into internationally in this Treaty to keep their territories forever free from nuclear weapons - will constitute a measure which will spare their peoples from the squandering of

TREATY: PROHIBITION OF NUCLEAR WEAPONS

their limited resources on nuclear armaments and will protect them against possible nuclear attacks on their territories, and will also constitute a significant contribution towards preventing the proliferation of nuclear weapons and a powerful factor for general and complete disarmament.

It is also worth noting that the Final Document approved by the special session of the UN General Assembly devoted to disarmament, which was held in 1978, contains several declaratory statements of a striking similarity to those included in the fourteen-year-old preamble of the Treaty of Tlatelolco.

The Latin American States, for instance, declared themselves convinced:

That the incalculable destructive power of nuclear weapons has made it imperative that the legal prohibition of war should be strictly observed in practice if the survival of civilisation and of mankind itself is to be assured.

That nuclear weapons, whose terrible effects are suffered, indiscriminately and inexorably, by military forces and civilian population alike, constitute, through the persistence of the radioactivity they release, an attack on the integrity of the human species and ultimately may even render the whole earth uninhabitable.

The United Nations, for its part, proclaimed in 1978 that:

Mankind today is confronted with an unprecedented threat of self-extinction arising from the massive and competitive accumulation of the most destructive weapons ever produced. Existing arsenals of nuclear weapons alone are more than sufficient to destroy all life on earth ...

Unless its avenues are closed, the continued arms race means a growing threat to international peace and security and even to the very survival of mankind ...

Nuclear weapons pose the greatest danger to mankind and to the survival of civilisation ...

Removing the threat of a world war - a nuclear war - is the most acute and urgent task of the present day. Mankind is confronted with a choice: we must halt the arms race and proceed to disarmament or face annihilation.

TREATY: PROHIBITION OF NUCLEAR WEAPONS

As to the articles of the treaty, their contents may be described in a nutshell as follows:
Article 1 defines the obligations of the parties. The four following articles (2-5) provide definitions of some terms employed in the treaty: contracting parties, territory, zone of application and nuclear weapons. Article 6 deals with the 'meeting of signatories', while Articles 7-11 establish the structure and procedures of the 'Agency for the Prohibition of Nuclear Weapons in Latin America' (OPANAL) created by the treaty, and state the functions and powers of its principal organs: the General Conference, the Council and the Secretariat. The five succeeding Articles (12-16) and paragraphs 2 and 3 of Article 18 describe the functioning of the 'control system', also established by the treaty. Article 17 contains general provisions on the use of nuclear energy for peaceful purposes, and Article 18 deals with nuclear explosions for the same purposes. Article 19 examines the relations of OPANAL with other international organisations, whereas Article 20 outlines the measures that the General Conforence shall take in cases of serious violations of the treaty, such measures consisting mainly of the simultaneous transmission of reports to the Security Council and the General Assembly of the United Nations. Article 21 safeguards the rights and obligations of the Parties under the Charter of the United Nations and, in the case of States members of the Organisation of American States, under existing regional treaties. Article 23 makes it binding for the contracting parties to notify the Secretariat of OPANAL of any international agreement concluded by any of them on matters with which the treaty is concerned. The settlement of controversies concerning the interpretation or application of the treaty is covered by Article 24.

Articles 22, 25-27 and 29-31 contain what is generally known as 'final clauses' dealing with questions such as privileges and immunities, signature, ratification and deposit, reservations (which the treaty does not admit), amendments, duration and denunciation, and authentic texts and registration. The transitional article specifies that 'denunciation of the declaration referred to in Article 28, paragraph 2, shall be subject to the same procedures as the denunciation' of the treaty, except that it will take effect on the date of delivery of the respective notification and not three months later as provided in Article 30, paragraph 2, for denunciation of the treaty. In paragraph 2 of Article 26, the Government of Mexico is designated the 'Depositary Government' of the treaty, whereas Article 7, paragraph 4, stipulates that the headquarters of OPANAL 'shall be in Mexico City'. Finally, Article 28 reflects in its text the compromise formula which, as already explained, overcame the most serious obstacle which confronted COPREDAL: the entry into force of the treaty.

TREATY: PROHIBITION OF NUCLEAR WEAPONS

As a complement to the preceding birds-eye view of the contents of the treaty, it seems advisable to examine more closely a few of its most significant provisions: those dealing with the obligations of the parties, the zone of application of the treaty, the definition of 'nuclear weapon', the system of verification and control and the use of nuclear energy for peaceful purposes. Some comments will also be in order with regard to the two additional protocols to the treaty.

As regards the obligations of the parties to the treaty, the Latin American States have drawn up a definition which is undoubtedly one of the most comprehensive ever produced on a world or regional level.

Under Article 1 of the treaty, the contracting parties undertake to 'use exclusively for peaceful purposes the nuclear material and facilities which are under their jurisdiction and to prohibit and prevent in their respective territories' both 'the testing, use, manufacture, production or acquisition by any means whatsoever of any nuclear weapons' and 'receipt, storage, installation, deployment and any form of possession of any nuclear weapons', by the parties themselves, directly or indirectly, on behalf of anyone else, by anyone on their behalf or in any other way.

The parties also undertake 'to refrain from engaging in, encouraging or authorising, directly or indirectly, or in any way participating in the testing, use, manufacture, production, possession or control of any nuclear weapon'.

The provisions of Article 4 of the treaty concerning its zone of application resulted from the procedure adopted in Article 28 for the entry into force of the treaty. This procedure had as a consequence two possible interpretations of the term 'zone of application': one entailing a moveable zone, in constant progression, and the other a fixed, clearly defined zone. These two different concepts are outlined, respectively, in paragraphs 1 and 2 of Article 4.

The first of these paragraphs, establishing that 'the zone of application of the Treaty is the whole of the territories for which the Treaty is in force', is that which has been used until the present, and, according to what was therein contemplated the extension and population of the zone has grown gradually as the number of contracting States has increased.

In the second paragraph it is stated that, 'upon fulfilment of the requirements of Article 28, paragraph 1, the zone of application of this Treaty shall also be that which is situated in the western hemisphere within the following limits'. Such limits are defined according to a series of geographical coordinates that can be easily consulted in the Treaty. It suffices to say that, on the one hand, a zone so defined includes considerable areas of the high seas which, in the western part of South America, extend to hundreds of

TREATY: PROHIBITION OF NUCLEAR WEAPONS

kilometres from the coasts, without naturally implying any pretension of sovereignty or jurisdiction over these sectors. Moreover, in light of the fact that the northern-most loxodromic line of the zone corresponds to 35 degrees north latitude, the paragraph in which it is explained expressly excepts the continental part of the territory of the United States and its territorial waters, which, had that not been so specified, would have been included in the Latin American nuclear-weapon-free zone, given that it reaches south of the parallel mentioned above.

The definition of the term <u>nuclear weapon</u>, which the Preparatory Commission finally approved after considering and rejecting several drafts, was included in Article 5 of the Treaty of Tlatelolco. It has the merit of being objective, precise and in accordance with the most recent technological advances. For the purposes of the treaty, 'a nuclear weapon is any device which is capable of releasing nuclear energy in an uncontrolled manner and which has a group of characteristics that are appropriate for use for warlike purposes'. In addition, the treaty provides that 'an instrument that may be used for the transport or propulsion of the device is not included in this definition if it is separable from the device and not an indivisible part thereof'.

As already mentioned, the provisions on <u>verification and control</u> appear in Articles 12-16 and Article 18, paragraphs 2 and 3. As UN Secretary-General U Thant emphasised in his message to the Preparatory Commission when the treaty was approved, on 12 February 1967, those provisions mark the first time that an international treaty dealing with disarmament measures includes an effective control system with its own permanent organs of supervision. The system calls for the full application of IAEA safeguards, but its scope is much greater. On the one hand it is to be used not only to verify 'that devices, services and facilities intended for peaceful uses of nuclear energy are not used in the testing or manufacture of nuclear weapons', but also to prevent any of the activities prohibited in Article 1 of the treaty from being carried out in the territory of the contracting parties with nuclear materials or weapons introduced from abroad, and to make sure that any explosions for peaceful purposes that might be carried out are compatible with Article 18 of the treaty. On the other hand, the treaty assigns important functions of control to the three main organs of OPANAL. Moreover, it also provides for the submission by the parties of periodic and special reports, for special inspections under certain circumstances, and for the transmission of the reports on those inspections to the UN Security Council and General Assembly.

Regarding the <u>use of nuclear energy for peaceful purposes</u>, from the beginning of the deliberations at the REUPRAL in November 1964, one of the fundamental concerns

TREATY: PROHIBITION OF NUCLEAR WEAPONS

of the participating States - as is shown by the fact that the first resolution adopted at that meeting applied to this question - was to spell out that, for the purposes they had in mind, 'denuclearisation' should be understood to mean the absence of nuclear weapons but not, of course, the rejection of the peaceful uses of the atom. On the contrary, in that very same resolution they emphasised the appropriateness of encouraging international cooperation in the peaceful uses of nuclear energy, particularly for the benefit of the developing countries.

Subsequently, the second and third sessions of the Preparatory Commission adopted similar texts which, with slight modifications, were to become one of the paragraphs in the preamble to the treaty, drafted in the following terms:

> ... The foregoing reasons, together with the traditional peace-loving outlook of Latin America, give rise to an inescapable necessity that nuclear energy should be used in that region exclusively for peaceful purposes, and that the Latin American countries should use their right to the greatest and most equitable possible access to this new source of energy in order to expedite the economic and social development of their peoples.

The treaty itself establishes the right, with no limitations other than those that may flow from the obligations assumed under the treaty, to use nuclear energy for peaceful purposes, and specifically provides, in Article 17, that:

> Nothing in the provisions of this Treaty shall prejudice the rights of the Contracting Parties, in conformity with this Treaty, to use nuclear energy for peaceful purposes, in particular for their economic development and social progress.

It was precisely for the purpose of avoiding any misunderstanding concerning the scope of the treaty and to indicate clearly that what was intended was not civil denuclearisation but only military denuclearisation, that the Preparatory Commission decided, at its last session, to change the original name of the treaty from 'Treaty for the Denuclearisation of Latin America' to 'Treaty for the Prohibition of Nuclear Weapons in Latin America'.

The desire to encourage and promote the peaceful utilisation of nuclear energy could not, however, have led the authors of the treaty to forget its primary object which is set forth in clear, precise and unambiguous terms in Article 1 of the instrument, by which the contracting parties undertake, <u>inter alia</u>, as already recalled, 'to refrain from engaging in, encouraging or authorising, directly or indirectly, or in any

TREATY: PROHIBITION OF NUCLEAR WEAPONS

way participating in the testing, use, manufacture, production, possession or control of any nuclear weapon'.

Thus, when drafting the provisions which would later be included in Article 18 dealing with nuclear explosions for peaceful purposes, special care was exercised to avoid any attempts to test or manufacture nuclear weapons under the pretext of carrying out such explosions for peaceful purposes, attempts which would completely negate the fundamental purpose involved, the very raison d'être of the treaty.

To this end, the first paragraph of Article 18 contains the provision that the contracting parties may carry out explosions of nuclear devices for peaceful purposes, but only if they can show that such explosions are feasible without violation of 'the provisions of this article and the other articles of the treaty, particularly Articles 1 and 5'. In the last analysis, this means that the explosions in question may be carried out directly by the parties to the treaty only if they do not require the use of a nuclear weapon as defined in Article 5 of the treaty.

An objective analysis of Article 18 shows therefore that its paragraph 1, as the text reads, is clearly subordinated to Articles 1 and 5 of the treaty. This means that for one of the contracting parties to carry out directly a peaceful nuclear explosion, it will have to prove previously that a nuclear weapon will not be required for that explosion; that is to say, in accordance with the objective definition contained in Article 5 of the treaty, that it will not require 'any device which is capable of releasing nuclear energy in an uncontrolled manner and which has a group of characteristics that are appropriate for use for warlike purposes'.

Since the consensus of the experts is that this is at present impossible, it must obviously be concluded that the States parties to the treaty will not be able to manufacture or acquire nuclear explosive devices, even though they may be intended for peaceful purposes, unless and until technological progress has developed, for such explosions, devices which cannot be used as nuclear weapons.

There is nothing in the treaty, however, that precludes the States parties to accept, as expressly provided in paragraph 4 of Article 18, 'the collaboration of third parties' - obviously meaning nuclear weapon States - for the purpose set forth in the first paragraph of the article, i.e. explosions for peaceful purposes, on the condition that they comply with the various obligations specified in paragraphs 2 and 3, and which relate to advance information and acceptance of measures of observation, verification and control to be carried out both by the General-Secretary and the Council of OPANAL, and by the IAEA.

The two Additional Protocols to the treaty have identical preambles. Their text recalls UN resolution 1911 (XVIII) and states the conviction that the treaty 'represents an important

TREATY: PROHIBITION OF NUCLEAR WEAPONS

step towards ensuring the non-proliferation of nuclear weapons'; points out that it 'is not an end in itself but, rather, a means of achieving general and complete disarmament at a later stage', and expresses the desire to contribute 'towards ending the armaments race'.

The operative parts of the protocols are naturally different from one another, although they have in common: identical duration (the same as that of the treaty); and entry into force for the States which ratify each Protocol (the date of the deposit of the respective instruments of ratification).

Under Article 1 of <u>Additional Protocol I</u>, those extra-continental States which, <u>de jure</u> or <u>de facto</u>, are internationally responsible for territories lying within the limits of the geographical zone established by the treaty would, upon becoming parties to the protocol, agree 'to undertake to apply the statute of denuclearisation in respect of warlike purposes as defined in Articles 1, 3, 5 and 13 of the Treaty' to such territories.

One aspect which should be borne in mind in connection with this protocol is the following: it does not give those States the right to participate in the General Conference or in the Council of the Latin American Agency. But neither does it impose on them any of the obligations relating to the system of control established in Article 14 providing for semi-annual reports, in Article 15 providing for special reports, and in Article 16 providing for special inspections. In addition, the prohibition of reservations included in the treaty's Article 27 is not applicable to the protocol. Thus, in the protocol the necessary balance has been preserved between rights and obligations: although the rights are less extensive, the obligations are also fewer.

With regard to <u>Additional Protocol II</u>, the obligations assumed by the nuclear powers parties to the protocol are stated in its Articles 1 through 3 in the following terms:

- Respecting 'in all its express aims and provisions' the statute of denuclearisation of Latin America in respect of warlike purposes, as defined, delimited and set forth' in the Treaty of Tlatelolco.

- Not contributing 'in any way to the performance of acts involving a violation of the obligations of Article 1 of the Treaty in the territories to which the Treaty applies'.

- Not using or threatening to use 'nuclear weapons against the contracting parties of the Treaty'.

CONCLUSIONS

The importance of nuclear-weapon-free zones has been emphasised several times by the United Nations. The General

TREATY: PROHIBITION OF NUCLEAR WEAPONS

Assembly in its resolution 3472 B(XXX) of 11 December 1975 stated that 'nuclear-weapon-free zones constitute one of the most effective means for preventing the proliferation, both horizontal and vertical, of nuclear weapons and for contributing to the elimination of the danger of a nuclear holocaust'.

Subsequently, in June 1978, the Assembly, in the Programme of Action adopted by consensus at its first special session devoted to disarmament, stressed the significance of the establishment of nuclear-weapon-free zones as a disarmament measure and proclaimed that 'the process of establishing such zones in different parts of the world should be encouraged with the ultimate objective of achieving a world entirely free of nuclear weapons'.

The weight which the international community attaches to the Latin American nuclear-weapon-free zone was manifest from the very moment when the Treaty of Tlatelolco was presented to the General Assembly, which, in its resolution 2286 (XXII) of 5 December 1967, welcomed it 'with particular satisfaction' and declared that it 'constitutes an event of historic significance in the efforts to prevent the proliferation of nuclear weapons and to promote international peace and security'. Such weight has been once again evidenced when, in the general debate of the Assembly's special session on disarmament, no less than forty-five States had enthusiastic comments for the treaty.

The Treaty of Tlatelolco has shown the crucial importance of ad hoc preparatory efforts - such as those carried out for two years by COPREDAL - in attaining the desired goal. Furthermore, the Latin American nuclear-weapon-free zone that is now nearing completion has become in several respects an example which, notwithstanding the different characteristics of each region, is rich in inspiration and profitable lessons for all States wishing to contribute to the broadening of the areas of the world from which those terrible instruments of mass destruction - nuclear weapons - would be forever proscribed.

NOTES

The author of this study presided over the 'Preliminary Session on the Denuclearisation of Latin America' (REUPRAL) held in 1964, as well as over the four sessions of the 'Preparatory Commission for the Denuclearisation of Latin America' (COPREDAL), held from 1965 to 1967, which made possible the elaboration and approval of the Treaty for the Prohibition of Nuclear Weapons in Latin America known as the Treaty of Tlatelolco. He was also Chairman of the 'Preliminary Meeting' (REOPANAL), contemplated in Article 28 (3) of the treaty and of the two parts of the first session of the

TREATY: PROHIBITION OF NUCLEAR WEAPONS

General Conference of the Agency for the Prohibition of Nuclear Weapons in Latin America (OPANAL) which took place in 1969 and 1970, respectively. In 1982 he was awarded the Nobel Peace Prize.

Chapter Four

THE TREATY OF RAROTONGA:
THE SOUTH PACIFIC NUCLEAR-FREE ZONE

Ramesh Thakur

The pursuit of nuclear non-proliferation has been a major international concern of our times. The Antarctic Treaty of 1959 is of great historical significance for having created the world's first nuclear-free zone (NFZ). Article 5 of the Treaty prohibits any nuclear explosions and the disposal of radio active waste in Antarctica. The Treaty of Tlatelolco of 1967 established the first internationally recognised Nuclear-Weapon-Free-Zone (NWFZ) in a populated region of the world, namely Latin America. The Non-Proliferation Treaty (NPT) of 1968 was an attempt to bring in a global regime to prevent the acquisition of nuclear weapons by non-nuclear-weapon states (NNWS). States in the latter category can adhere to the NPT while accepting a stationing of nuclear weapons on their territories, as long as they do not exercise jurisdiction and control over the weapons. West Germany is an obvious example of such a country. A NWFZ, however, prohibits such stationing of nuclear weapons. The three essential characteristics of NWFZ are non-possession, non-deployment and non-use of nuclear weapons.

NWFZs can help to strengthen and promote non-proliferation by providing a means of extending and reinforcing the NPT. In fact Article 7 of the latter accepts that 'Nothing in this Treaty affects the right of any group of states to conclude regional treaties in order to assure the total absence of nuclear weapons in their respective territories'. The article merely acknowledged that one such treaty had been negotiated more or less simultaneously with the NPT. The second NWFZ in an inhabited region was not to be established for another eighteen years. At the sixteenth South Pacific Forum meeting held in Rarotonga, Cook Islands, Forum countries adopted the South Pacific Nuclear-Free Zone Treaty on 6 August 1985 (Hiroshima Day). The preamble to the treaty expresses the commitment to world peace, a grave concern at the continuing nuclear arms race, the conviction that every country bears an obligation to strive for the elimination of nuclear weapons, a belief in the efficacy of regional arms control measures, and a

THE TREATY OF RAROTONGA

reaffirmation of the NPT for halting nuclear proliferation. The core NFZ obligations are contained in Articles 3-7. Each party agrees not to manufacture or otherwise acquire, possess or have control - or seek to do so - over any nuclear device, not to assist or encourage others to make or acquire nuclear weapons; to prevent the stationing or testing of nuclear weapons on its territory; not to dump radioactive wastes at sea anywhere in the zone, and to prevent such dumping by others in its territorial sea.

Discussions at the United Nations had, by the mid-1970s, identified nine major principles as the guiding elements of a NWFZ:

1. the initiative should come from the countries of the region;
2. the specific provisions of the NWFZ should be negotiated between the regional member states in the form of a multilateral treaty establishing the zone in perpetuity;
3. while adherence to the treaty should be voluntary, the NWFZ must nevertheless embrace all militarily significant states in the region;
4. existing treaty relationships within the zone should not be disturbed;
5. there should be an effective verification system;
6. peaceful nuclear development should be allowed;
7. the zone should have clearly defined and recognised boundaries;
8. in defining the territory of the zone, members must respect international law, including freedom of the high seas and straits used for international navigation and of international airspace; and
9. the NWFZ should have the support of the nuclear-weapon states.

The United Nations group mirrored international reality in being divided on some major issues confronting NWFZ proposals. 'Nuclear weapons' were not defined; it was not clear whether negative security guarantees by nuclear powers (i.e., undertakings to refrain from using nuclear devices against NNWS) were an essential or merely a desirable condition; the line between peaceful nuclear development and peaceful nuclear explosions remained blurred; the relationship between a NWFZ and the NPT can be contentious; and the issue of existing security relationships between NWFZ member-states and outside powers is even more troublesome. Thus the NWFZ concept in established United Nations vocabulary does not prohibit the temporary presence of nuclear vessels during transit or on port calls, and it does not necessarily preclude the acquisition of sensitive nuclear facilities and material tantamount to having a nuclear-weapon capability or producing untested nuclear bomb components. Nevertheless, it will

THE TREATY OF RAROTONGA

be useful to follow the UN criteria in order to examine the nature and implications of the South Pacific zone.

REGIONAL INITIATIVE

The initiative for the SPNFZ not only came from within the region; in fact it has a considerable history behind it. The Australian Labour Party (ALP) was attracted to the concept as early as 1962, after the successful denuclearisation of Antarctica in 1959. The New Zealand Labour Party (NZLP) leader Norman Kirk was drawn to the idea after the conclusion of the Latin American zone in 1967. Between 1972-75, the Labour government of NZ pursued a strongly anti-nuclear stance in its foreign policy, banning port visits by nuclear-armed ships, opposing French nuclear testing in the South Pacific, supporting calls for an Indian Ocean zone of peace, and laying the groundwork for a regional NFZ.

The most appropriate agency for pursuing a regional initiative is the South Pacific Forum. At the 1975 meeting in Tonga, Forum countries unanimously commended the idea of establishing a SPNFZ as a means of keeping the region free of the risk of nuclear contamination. The Forum communiqué was issued on 3 July 1975. On 15 August, Fiji and New Zealand wrote to the UN Secretary-General asking that the item of a South Pacific NWFZ be included on the agenda of the forthcoming 30th session of the General Assembly. The explanatory memorandum and draft resolution accompanying the note emphasised that regional approaches to eliminating nuclear weapons had become important because of the meagre results obtained by global efforts. Denuclearisation provisions of the Antarctic Treaty and Tlatelolco were explicitly recalled, as also the declaration by the NPT Review conference endorsing internationally recognised NWFZs. In this context, and with the aim of enhancing the security and welfare of South Pacific peoples and minimising risks to their health and environment the operative clauses of the draft resolution endorsed the idea of establishing a NWFZ in the South Pacific in cooperation with the nuclear powers and with the possible assistance of the UN Secretary-General [1].

When the time came to present the draft resolution in the General Assembly, New Zealand and Fiji were joined by Papua New Guinea as co-sponsor. Having but just achieved independence on 16 September 1975 and been admitted to UN membership, Papua New Guinea's first political act at the United Nations was to co-sponsor the NWFZ proposal. The resolution was adopted in the first committee on 28 November by a vote of 94-0, with 18 abstentions; and in the General Assembly on 11 December by a vote of 110-0-20 (Resolution 3477). China was the only nuclear power to vote in favour. Britain, France, the USA and the USSR expressed general sympathy,

THE TREATY OF RAROTONGA

but abstained on the vote because of fears for their rights upon the high seas [2].

Labour's hold on office proved too short-lived to develop a South Pacific NWFZ from an idea to a programme of action. The National Party government of Robert Muldoon (1975-84) was perceived as unsympathetic to such notions. Malcolm Templeton, who was the NZ Permanent Representative to the UN at the time, has revealed that the new National government abandoned the NWFZ initiative, leaving him 'stuck as the leader of the New Zealand delegation with a proposal which my government no longer supported'. As he noted,'this was not one of my more pleasant diplomatic experiences' [3]. In February 1976, Muldoon described his attitude as being more realistic in accepting that without assurances of respect from the nuclear powers, a mere declaration of a NWFZ would remain an empty gesture. In March 1976, Australia and New Zealand used the South Pacific Forum meeting in Rotorua to obtain a joint declaration that a regional NWFZ would respect the freedom of the high seas and the existing security arrangements. In June, Muldoon publicly quoted a defence memorandum addressed to the former Labour Government expressing disquiet over testing the ANZUS relationship through NWFZ initiatives. The return of American nuclear ships to New Zealand harbours was entirely consistent with National's approach.

The conservative governments of Australia and NZ can fairly be described as having pursued a SPNFZ for the purposes of establishing an alibi rather than achieving genuine progress. The political environment changed dramatically with the return of labour governments in the two countries. In Australia, the ALP government announced its intention to promote a SPNFZ shortly after assuming office in March 1983. It took its proposal to the 14th South Pacific Forum meeting held in Canberra in August 1983. The Forum countries reiterated their strongest protests and condemnations of continued French nuclear testing in the South Pacific, and expressed strong opposition to proposals for dumping and storing nuclear waste material in the Pacific. The Australian initiative in reviving the NWFZ concept was commended, but more time was considered necessary to study its implications and engage in discussion [4].

The Australians took their proposal a step further at the South Pacific Forum meeting in Tuvalu in August 1984. The Forum decided to establish a working party of officials to look at the detailed issues involved in actually establishing a treaty-based NWFZ; the group was to examine the whole gamut of legal, political and physical problems, and report back to the Forum meeting in the Cook Islands in August 1985. Calls for its formation arose from the feeling that talks on the topic had gone on for long enough; it was time to examine the practical issues involved in creating a zone which

THE TREATY OF RAROTONGA

would prohibit the manufacture, testing, use, storage and acquisition of nuclear weapons, and the dumping of nuclear waste [5]. The working party had five meetings to negotiate a draft treaty: in Suva in November 1984, and in Canberra, Wellington, Suva in February, April, May and June 1985 respectively. Various Forum countries participated in the working party sessions; others were kept informed of its deliberations in order to facilitate the continuing discussions. The Federated States of Micronesia also attended as an observer. While the working party identified the main elements of a SPNFZ treaty, the actual drafting was entrusted to a legal sub-committee set up at the Suva meeting in November 1984. The draft treaty was presented to and adopted by the full Forum meeting to the Cook Islands in August 1985.

NFZ TREATY

The goal of a SPNFZ thus clearly originated in the region, was pursued through regional channels, and was adopted by the regional Forum which encompasses all the countries of the region. The zone was established also in the form of Forum countries adopting a formal treaty.

MEMBERSHIP

The voluntary adherence of all militarily significant states has not been an unrealistic goal for the South Pacific region. The difficulty was how best to convert support for a NFZ in the abstract into signature of a NFZ treaty in the particular. Forum countries had different approaches in respect of the content of the treaty. The conservative pole, represented by Fiji and Tonga, was determined to reject any concrete impact upon existing security policies and practices; they would have preferred to leave open the option of an American stationing of nuclear weapons in the region at a future date. Tonga's opposition to the concept of a NFZ was strong enough that it did not participate in the working party sessions leading up to the Rarotonga meeting. The radical pole, represented by Vanuatu, favoured banning all types of nuclear involvement in the region and extending the zonal boundary to bring Micronesia within its scope. Papua New Guinea and the Solomon Islands too leaned towards a more comprehensive treaty which banned the transit of nuclear ships through the region. NZ, although more radical in its national policy, was prepared to support the Australian draft as a compromise broadly acceptable to the Forum countries. The Hawke government for its part insisted that ANZUS obligations and its security relationship with the US were not negotiable.

THE TREATY OF RAROTONGA

The Treaty of Rarotonga was signed on 6 August 1985 by Australia, the Cook Islands, Fiji, Kiribati, New Zealand, Niue, Tuvalu and Western Samoa. Papua New Guinea signed the treaty on 16 September, the tenth anniversary of its independence. Vanuatu, however, broke the Forum tradition of consensual decisions when Prime Minister Walter Lini declared on 7 August (the day after the treaty's adoption) that he would not be signing the treaty. Nauru became the tenth signatory in July 1986. (The remaining two Forum members are the Solomon Islands and Tonga.) The zone will enter into force when ratified by eight countries; for countries ratifying the treaty thereafter, the treaty will enter into force on the date of deposit of its instrument of ratification. The Director of the South Pacific Bureau for Economic Cooperation (SPEC) is designated as the depository of the treaty, its protocols, and instruments of ratification. Only Forum members can accede to the SPNFZ treaty. While the duration of the treaty is indefinite, every party is given the right to withdraw in the event of a core violation of treaty provisions by any member; withdrawal can be effected upon a twelve month notice to the Director of SPEC.

EXISTING SECURITY ARRANGEMENTS

As the preceding section hinted, Forum countries' security perceptions and policies are diverse rather than uniform. Australia hosts critical command and guidance facilities for the US strategic deterrent forces. Australia, Fiji and Tonga also accept port calls by US ships within the global US policy of neither confirming nor denying the presence of nuclear armaments aboard vessels. NZ, Vanuatu and now even the Solomons reject port visits by all ships unless satisfied that they are neither nuclear-propelled nor carrying nuclear weapons. A major stumbling block to achieving a NWFZ in any given region is a potential clash with the principle of undiminished security for everyone, including all countries in the region and the major powers external to the region.

Existing security arrangements adopted and maintained by member countries have consistently been the starting point for all Forum NFZ initiatives, including that of 1975. The Forum communique in regard to the 1983 Australian initiative similarly reaffirmed 'the sovereign right of Governments to make their own decisions on their alliance and defence requirements, including access to their ports and airfields by the vessels and aircraft of other countries' [6]. In setting up a working party of officials, the 1984 Forum reiterated the same principle. It is not surprising, therefore, that nothing in the Treaty of Rarotonga infringes the unqualified sovereign rights of parties to determine their own security arrangements short of nuclear weapons acquisition or stationing. In fact the

THE TREATY OF RAROTONGA

treaty specifically states the right of each party 'to decide for itself whether to allow visits ... transit ... and navigation' by foreign ships or aircraft to its territorial air and sea space (Article 5.2). The adoption of a NFZ treaty was certainly made easier by the fact that the South Pacific is already a de facto NFZ. With the accession of Kiribati to the NPT in May 1985, Vanuatu is the only internationally independent South Pacific country not to figure among the 130 parties to the NPT by mid-1985; Vanuatu's parliament adopted a resolution in 1982 comprehensively banning anything and everything nuclear from all its territory. Nuclear weapons are not deployed in any South Pacific country, and there is neither the intention to acquire nuclear weapons nor the wish to station foreign nuclear weapons in the region. The value of the Treaty of Rarotonga lies, therefore, primarily in formalising the existing situation and precluding changes to the contrary. It is probably worth noting also that nuclear technology cannot simply be disinvented. Consequently, nuclear weapons are going to be an inescapable part of the world's political reality for the foreseeable future, and the non-nuclear countries have to learn how best to cope with the possession of weapons of mass destruction by some countries.

VERIFICATION

The Treaty of Rarotonga establishes two levels of control machinery to verify compliance of parties with treaty obligations: regional and international. Regionally, the control system comprises the three elements of reports and information exchange, consultations, and a complaints procedure. Any significant development affecting the operation of the treaty has to be reported to the Director of SPEC, who then circulates the reports to all members; the parties are also encouraged to exchange pertinent information through the Director; and the Director is required to submit an annual report to the Forum on the status of the treaty. The Director is further authorised to convene an ad hoc Consultative Committee at the request of any party for consultation and cooperation on any matter in connection with the treaty. Each party is entitled to appoint one representative to the Committee; a quorum is constituted by half the parties; and decisions are to be made by consensus or two-thirds majority voting. A party suspecting violations of the treaty by any other party must first raise the issue with the latter. If not satisfied with the response, the complainant can request a meeting of the Consultative Committee, furnishing evidence as appropriate. If the matter is not explained to the satisfaction of the Committee, it is empowered to direct that a special inspection be carried out by three qualified inspectors appointed in consultation with the complainant and 'defendant'

THE TREATY OF RAROTONGA

parties. The inspectors are to be given 'full and free access to all information and places' within the territory of any party; they will operate under the direction of, and shall report in writing to, the Consultative Committee. The Committee then reports to the Forum members on whether a party is in breach of treaty obligations; if the conclusion does establish such a breach, then the subject is to be referred to a promptly convened meeting of the South Pacific Forum.

At a regional level, therefore, the Treaty of Rarotonga emulates the Latin American model in establishing bureaucratic machinery to facilitate on-site challenge and spot inspections. Internationally, the SPNFZ requires submission to full-scope IAEA safeguards. The International Atomic Energy Agency (IAEA), although autonomous, is a member of the UN system and reports annually to the General Assembly on its work. One of its major functions is to apply safeguards to ensure that nuclear materials and equipment intended for peaceful uses are not diverted to military uses. The safeguards system, which constitutes the international community's first attempt to establish a control system over an industry of strategic importance, includes four main steps:

1. IAEA experts examine the design of a state's nuclear plants to check that it permits effective control;
2. the state is required to keep detailed records of plant operations and flow and inventory of new materials;
3. the government concerned supplies periodic reports to the IAEA based on those records; and
4. the IAEA sends inspectors for spot-checks in nuclear plants (inspectors for each country are designated with the consent of that country).

The Treaty of Rarotonga requires IAEA safeguards to be applied on all sources of special fissionable material in all peaceful nuclear activities within the territory of a party, under its jurisdiction, or carried out under its control anywhere. The goal of the safeguards system is to verify the non-diversion of nuclear material from peaceful nuclear activities to nuclear explosive devices. Each party is required to negotiate, conclude and bring into force IAEA safeguards within 18 months of the SPNFZ being in force for it. Each party is given the right to request copies of the overall conclusions of all IAEA inspection reports on the nuclear activities of zonal members.

With such a compliance and control system in force, applied to the complete nuclear fuel cycle for each party to ensure that any diversion of fissile material would be detected in good time, regional states as well as outsiders should have reasonable assurance that the NFZ status is not being violated in the South Pacific.

THE TREATY OF RAROTONGA

PEACEFUL NUCLEAR DEVELOPMENT

As expected, the Treaty of Rarotonga does not prohibit peaceful nuclear activities. However, member parties are forbidden to transfer material or equipment, even for peaceful purposes, to NNWS not subject to NPT safeguards, or to NWS not subject to IAEA safeguards. That is, parties are enjoined to ensure that any transfer of nuclear technology or material conforms to strict non-proliferation measures in order to provide assurance of exclusively peaceful use.

GEOGRAPHIC SCOPE

The boundaries of the SPNFZ generally follow the territorial limits of the South Pacific Forum members. To the east, the SPNFZ is joined with the Latin American zone at 115°W; its southern boundary coincides with the outer limit of the Antarctic Treaty at 60°S; its western boundary is formed by Australia's and Papua New Guinea's western borders; and its northern boundary is formed generally by the equator, with a few ups and downs to fit certain specified contours. Essentially, therefore, the SPNFZ ranges from 115°E to 115°W, and from the equator to 60°S. That is a considerable area. Indeed, when we examine the zone alongside the adjoining Latin American and Antarctic zones, then Africa and Indonesia are the only areas in the entire southern hemisphere not within the ambit of a NWFZ.

FREEDOM OF THE SEA

The eighth criterion demands respect for the law of the high seas and straits. One fundamental difference between a SPNFZ and all other zonal proposals in populated regions of the world is the balance between territorial lands and international waterways. For example, the right to free passage on the high seas does not seriously detract from the significance of the Latin American NWFZ, since the zone is designed primarily to encompass the continental land mass. In the South Pacific, by contrast, territorial lands are typically specks in an ocean vastness. A NFZ which does not extend to the high seas inevitably therefore degrades the value of a South Pacific zone when compared to Latin America.

Attempts to include the high seas within a NFZ scope however would create more problems than it would solve. To begin with, verification is a greater problem in this context than if a NFZ was territorially limited. But it is not an insuperable problem, for it is largely technical. Ocean floor electronic sonar technology already in use in the Atlantic could presumably be adapted for monitoring and verification of a South Pacific NFZ. Or if not present-day technology, then at least one developed in the future would serve the purpose.

A somewhat greater problem would lie in the fact that detection technology either lies or will lie with those very states with the capability to violate NFZ status on and under the high seas. If verification technology is possible, then so presumably is electronic jamming or confusing technology for those nuclear powers determined to sail in South Pacific seas.

The most serious of the problems confronting a NFZ which applies to the high seas are not technical ones. Rather, they are legal-political. The law of the high seas originated in the rule that the high seas are not open to acquisition by states individually or collectively. Because it is a policy concept or a general principle of international law, its application to specific problems is often vague and imprecise. Nevertheless, there is consensus that the general concept embodies four freedoms, namely freedom of navigation; freedom of fishing; freedom to lay submarine cables and pipelines; and freedom to fly over the high seas. All four freedoms may be exercised in time of war and armed conflict as well as in times of peace [7].

Imposing restrictions on established maritime rights would not be proper because ultimately it would be damaging to the goal of creating a civilised world ruled by law and reason. The South Pacific countries would not benefit the cause of world peace by violating international law; they should rather be striving to strengthen it at points where the actions of others have weakened it. Forcible restrictions would be unrealistic, secondly, because of the impossibility of enforcement. It is difficult enough to enforce the accepted law against the determined opposition of a major power. No South Pacific country individually, nor all of them collectively, could enforce an illegal ban on nuclear ships in the international waters of the South Pacific.

It is also necessary to distinguish between abhorrence of nuclear wars and opposition to nuclear weapons. Deterrence depends upon the guaranteed vulnerability of industrial-population targets to enemy attack, and survivability of nuclear weapons against enemy attack. In the strategic triad of land-based ICBMs, bombers and submarines, the last are the least vulnerable, and therefore the most stabilising element in the nuclear equation. Progressively expanding NFZs that included high seas in their geographic coverage would undoubtedly reduce areas of nuclear weapons deployment. But they would also destabilise the US-Soviet nuclear deterrence, and so could make the outbreak of nuclear war more likely, not less. It is possible, therefore, simultaneously to reject nuclear war fighting doctrines, accept the need for a rational policy of nuclear deterrence, and accept the advantages of

THE TREATY OF RAROTONGA

sea-based missiles in injecting stability into the nuclear balance. In the 1975 South Pacific NWFZ proposal at the UN, Fiji and New Zealand had accepted that imposing restrictions on the maritime rights of unwilling third parties was neither proper nor realistic. What they hoped for instead was that once a NWFZ had been concluded, and its relationship with nuclear powers defined, then it might be possible to limit the deployment of nuclear weapons in ways which did not offend the earnest wishes of South Pacific peoples. The position is much the same with the NFZ adopted in 1985. The Forum had emphasised the principle in its communiques in 1983 and 1984 when taking up the NFZ proposal for consideration; the Treaty of Rarotonga specifically states that 'nothing' in it 'shall prejudice or in any way affect the rights, or the exercise of the rights, of any State under international law with regard to freedom of the seas' (Article 2.2).

NUCLEAR POWERS SUPPORT

The final criterion merely recognises the fact that a NFZ which does not obtain the endorsement of NWS is practically worthless. In 1975, NZ had been quite open to discussions with the NWS on how all reasonable preconditions attaching to their support could be put into effect. The Treaty of Rarotonga adopts the simple expedient, following Tlatelolco, of containing additional protocols for integrating NWS into the SPNFZ. Protocol 1 is addressed to France, the UK and the USA, and invites them to apply NFZ prohibitions on manufacture, stationing and testing, to their territories within the zone. Protocol 2 is addressed to the five NWS and contains the negative security guarantees. In it, each NWS party agrees not to violate the NFZ treaty, and not to use or threaten to use nuclear weapons against any treaty party, or in the territory of any party to the treaty or to Protocol 1. Protocol 3, which does not have a counterpart in Tlatelolco, prohibits the testing of any nuclear device anywhere in the region. A team of officials from Australia, the Cook Islands, Fiji, PNG and the Solomon Islands visited Beijing, Paris, Moscow, London and Washington in January-February 1986 to explain the SPNFZ to the five NWS and to seek their views on the draft protocols. The officials reported back to a meeting of the South Pacific Forum's Working Group on the SPNFZ in Suva in April, which in turn led to the approval of certain adjustments to the draft protocols by the full Forum meeting in Suva in August. For example, it was agreed to allow NWS to withdraw from their protocol obligations after three months' notice. NZ Prime Minister David Lange said that Britain and America had indicated that they would sign the protocols if the changes were effected [8].

THE TREATY OF RAROTONGA

In the abortive 1975 initiative, China had been the only NWS to support the NWFZ resolution without qualification. Similarly, it did not take China long in 1985 to indicate support for the Treaty of Rarotonga, and a willingness to accede to Protocol 2. British accession to the three Protocols can safely be presumed as well because of the care taken not to infringe upon international rights of navigation. John Stanley, the British Minister for Armed Forces, described the NZ ban on nuclear ship port calls as a setback for Pacific security, but did not reject the SPNFZ. He merely said that an early British decision on endorsing the protocols was unlikely [9].

The Soviet Union has been a longstanding advocate of NWFZs. It may be unhappy with a SPNFZ which preserves the region as a de facto Western lake, permits the operation of nuclear weapons related facilities in Australia and US port visits, while denying transit facilities to any Soviet warship. But in the end the USSR did join a similar Tlatelolco regime, and can be expected to join the SPNFZ. One Soviet commentator noted that the South Pacific is 'part of the catchment area of the US Seventh Fleet equipped with Trident submarines and Tomahawk cruise missiles with nuclear warheads'. Noting right-wing US efforts to neutralise the treaty by propagating the myth of the Soviet threat, he attributed the treaty's adoption to growing concern at US militarisation and nuclearisation of the Pacific, as well as to opposition to French nuclear activity in the region. Prime Minister Hawke, by pushing the Washington line on transit rights and port calls, had 'weakened the efficacy of the treaty'. Soviet leader Mikhail Gorbachev had nevertheless welcomed efforts at creating a SPNFZ as 'praiseworthy' [10]. In a letter to the Prime Minister of the Cook Islands Tom Davis, who had hosted the 1985 Forum meeting which adopted the SPNFZ, Soviet Foreign Minister Eduard Shevardnadze declared in March 1986 that the USSR would sign the protocols if other nuclear powers did the same [11]. The Minister sought to exploit regional sentiment by praising the Forum's action for demonstrating that public statements of the wish to prevent a nuclear war were no longer enough. Soviet Deputy Foreign Minister Mikhail Kapitsa confirmed during a visit to NZ in August 1986 that the USSR would sign the SPNFZ protocols once the Treaty had been brought into force by eight Forum members' ratifications [12].

This leaves the US and France as the two problematical nuclear powers. In September 1985, Bob Hawke had approached the US to exert pressure on France to cease nuclear testing in the Pacific. The Reagan administration however confirmed the primacy of the North Atlantic over the South Pacific in its strategic worldview. The State Department described French nuclear tests as 'essential to the modernisation of the French nuclear deterrent'; in any case, the US

THE TREATY OF RAROTONGA

regarded 'these matters as French decisions' [13]. The Americans have been less than enthusiastic supporters of a SPNFZ. With nuclear proliferation not being an issue in the region, and with the Soviet profile barely visible even with the aid of powerful technological gadgetry, Washington does not see any of its major security goals being served by the Treaty of Rarotonga. On the other hand, if the Treaty helps to appease the strong anti-nuclear sentiment in the region without undermining US strategic position, then Washington would be better advised to settle for the moderate, Australian-initiated zone than to inflame anti-Americanism by unilateral dissociation. The final text of the treaty did not limit the movements and flexibility of US nuclear forces in the South Pacific; in fact the treaty is so innocuous that US officials joked about awaiting a 'message in a bottle' from the Cook Islands before deciding upon a formal response. The crack indicates relief in Washington rather than insensitivity towards the South Pacific peoples. It was reported in April 1986 that the US has adopted a tough uncompromising stand towards the Treaty of Rarotonga. Washington was reported to have concluded that the Australian initiative had been meant as a sop to ALP activists rather than as a serious commitment to regional nuclear arms control; that the SPNFZ would directly benefit Soviet global objectives to the detriment of US interests; and that the SPNFZ could create a serious precedent for the establishment of interlocking NWFZs which would impede the free movement of US forces around the world [14]. Senator Gareth Evans denied in the Australian Parliament on 6 May 1986 that the US held any misapprehension over Australian commitment to the SPNFZ, although some agencies within the US administration were sceptical about the whole concept of nuclear-free zones. The National Times thesis was also undermined by Richard Lugar during a visit to NZ in August. The influential chairman of the Senate Foreign Relations Committee thought that the protocols would be acceptable to the US 'in due course', and that the present 'bias' was in the direction of signing them. A decision is expected by the end of the year [15]. The Department of State is generally in favour of signing, although the European bureau sympathises with French opposition to the SPNFZ; but the Department of Defence is reluctant to see the US sign the protocols: the decision may have to rest with the President.

The Americans could afford to laugh at the Treaty of Rarotonga; not so the French. As Australian defence minister Kim Beazley conceded subsequently, the treaty was aimed primarily at France [16]. The political fallout from the French sinking of the Greenpeace boat Rainbow Warrior in Auckland harbour on 10 July 1985, killing one crew member, focused international attention on French nuclear testing in the region [17]. The fact that the French testing brought the first act of international terrorism to NZ also strengthened regional

THE TREATY OF RAROTONGA

opposition to French nuclear activities. The French programme of testing in the South Pacific began with a 25-30kt atmospheric explosion on 2 July 1966. There were 41 atmospheric tests in all, culminating with a blast on September 1974. Australia and NZ terminated atmospheric testing with a successful action against France at the World Court; the series of underground testing began with a 5kt explosion at Fangataufa Island on 5 June 1975. France exploded a 150kt device at Mururoa Atoll on 9 May 1985, the sixty-ninth in the underground series. It is worth noting that the United States and the Soviet Union signed a Threshold Test Ban Treaty in 1974 which prohibits underground tests of more than 150kt yield. The first test in 1986 was conducted on 26 April, and had an estimated yield of 4kt; tests of 3kt, 2kt and 20kt followed in May. The next series, following established pattern, can be expected in October-December.

Even after the international embarrassment of the Rainbow Warrior affair, therefore, France refused to budge on the issue of nuclear testing at Mururoa. Indeed President Francois Mitterrand flew to Mururoa in mid-September 1985 to emphasise French determination on the subject, an act that was interpreted - and probably intended - as a calculated rebuff to pressure from the South Pacific countries to cease testing in the region. Although France has so far failed to ratify Protocol 1 to the Treaty of Tlatelolco (which commits metropolitan countries to apply the denuclearisation provisions on emplacement and storage to their territories within the Latin American zone), the South Pacific is the only region in the world where nuclear explosive testing is presently carried out outside metropolitan territories. Not surprisingly, the French view the SPNFZ - again, probably correctly - as a means primarily of bringing further pressure to bear on their nuclear programme. French opposition to the SPNFZ is expressed on three counts: the major part of the zone comprises international seaspace free of any constraints; risks of horizontal proliferation are minimal in the South Pacific; the five NWS have already given negative security guarantees at an international level [18]. Given such an attitude on the part of the French government, it would appear unrealistic to expect a modification of their nuclear activity in the foreseeable future. The South Pacific countries simply lack the power of military might, diplomatic sanction or economic resources to compel France to terminate its testing programme. Such a conclusion is unlikely to stop regional countries from trying to change French policy, since in the long term the basis of opposition to French testing is more durable.

THE TREATY OF RAROTONGA

EVALUATION

The Third Review Conference of the NPT in its final declaration of September 1985 welcomed the achievement of the SPNFZ as being consistent with Article 7 of the NPT. Similarly, the 46 Commonwealth Heads of Government, meeting in the Bahamas in October 1985, welcomed the SPNFZ Treaty as an important step in global and regional efforts to prevent nuclear proliferation. The Treaty of Rarotonga in fact goes beyond the minimalist position for NWFZs as set out by the United Nations. Firstly, the title itself is of some significance. A NWFZ indicates an arms control objective. Because the South Pacific countries have been just as firmly committed to environmental goals, the treaty also prohibits the dumping of any nuclear waste in the zone (Article 7). In this respect the Rarotonga Treaty is an advance upon that of Tlatelolco, and more closely resembles the Antarctic Treaty. The legal implications of the zone for other countries are unclear, but its political message is not: Forum countries are united in opposition to 'nukeypooh' being dumped anywhere in their region [19]. The Treaty also calls for parties to support the conclusion of a global convention on the matter. The United Nations set up the London Dumping Convention in 1983 in an effort to control the dumping of radioactive waste at sea. In 1983, the Convention had imposed a two-year moratorium on low-level radioactive waste. At the 1985 meeting, Australia, Kiribati, Nauru and NZ from the South Pacific were among the 25 countries deciding to impose an indefinite moratorium. (Seven, including Japan, abstained; six, including Britain, France and the USA, voted against.) The resolution will become legally binding if it is proposed as an amendment to the Convention and adopted by a two-thirds majority; Kiribati and Nauru have indicated that they will so propose at the next meeting in October 1986 [20]. At the Forum meeting in August 1986, it was also agreed to negotiate a regime with China, France, Japan and the US to prohibit the dumping of nuclear waste. The talks would take place within the South Pacific Regional Environmental Programme (SPREP), and would include the question of the environmental effect of nuclear testing. Aimed at covering the Exclusive Economic Zones (EEZs) of Forum countries, SPREP also calls France's bluff in respect of nuclear testing at Mururoa. If, as France claims, the testing poses no attendant environmental risk, then there is no reason for France not to join the SPREP regime and apply its restrictions to the disposal of waste from Mururoa. Talks with the four Pacific nuclear energy powers are scheduled to start in November 1986.

The Treaty of Rarotonga is an improvement upon Tlatelolco also in its arms control objective. The Tlatelolco regime permits parties to 'carry out explosions of nuclear devices for peaceful purposes - including explosions which

THE TREATY OF RAROTONGA

involve devices similar to those used in nuclear weapons' subject to certain informational requirements (Article 18). The SPNFZ prohibits the manufacture, acquisition, control or testing 'of any nuclear explosive device' (Articles 3 and 6). Article 1 (c) defines nuclear explosive device as 'any nuclear weapon or other explosive device capable of releasing nuclear energy, irrespective of the purpose for which it could be used'. That is, 'peaceful nuclear explosions,' like India's in 1974, are prohibited in the South Pacific, but permitted in Latin America.

Rarotonga is an improvement upon Tlatelolco in its entry into force provisions. Article 28 of Tlatelolco stipulates three conditions for the zone to be established: all Latin American republics must sign and ratify; all relevant states must sign and ratify the two Protocols; IAEA safeguard agreements must be concluded. However, a party can waive these conditions, in which case the treaty comes into force for the party upon its ratification. As of 1984, 23 of the 33 eligible countries had exercised the waiver option and brought the treaty into force for themselves. The exceptions include four significant countries: Cuba, which has not signed; Argentina, which has signed but not ratified; Brazil and Chile, which have ratified without exercising the waiver. As noted earlier, France is yet to ratify Protocol 1. The Treaty of Rarotonga is considerably simpler - and therefore stronger - for going into force as soon as any eight parties have ratified; it goes into force for subsequent adherents upon their individual dates of ratification. The zone came into effect on 11 December 1986, when Australia became the eigth country to deposit its instruments of ratification with SPEC [21]. The seven countries which had already ratified were the Cook Islands, Fiji, Kiribati, New Zealand, Niue, Tuvalu and Western Samoa.

The South Pacific regime is stronger than the Latin American precedent also in the related sense of supplementing the NPT. Argentina, Brazil and Chile have all rejected the NPT as discriminatory; the only South Pacific country which does not adhere to the NPT, Vanuatu, is the most radically anti-nuclear country in the region. Rarotonga is an improvement upon Tlatelolco, finally, in introducing a longer delay before the regime can be dismantled. The Treaty of Tlatelolco imposes a three month period of denunciation (Article 30); Rarotonga requires a twelve month notice of withdrawal (Article 13).

Interestingly, the SPNFZ may strengthen non-proliferation goals somewhat unexpectedly even in respect of one individual country. As a non-Forum country, Indonesia is not a party to the Treaty of Rarotonga. Yet Indonesia is one of the few states whose military presence impinges on the South Pacific region. Moreover, Australians believe that the only real if remote local threat to Australian security is posed by Indonesia. This is because of Indonesia's size, its previous

THE TREATY OF RAROTONGA

willingness to use force to pursue major objectives, and its potential for conflict with Papua New Guinea which would have ramifications for Australian strategic interests. Moreover, the acquisition of nuclear weapons by Indonesia cannot be ruled out, despite its being a signatory to the NPT. We also know that the Australian defence establishment did not wish to forego the nuclear option as a response to another country in the region acquiring nuclear weapons. The Australian Defence Committee wanted to monitor regional nuclear programmes 'in order to ensure that the lead time for Australia could be matched with developments in other countries' [22]. Since Indonesia is the only other country in the region capable of nuclear weapons acquisition in the foreseeable future, its non-inclusion in a South Pacific NWFZ might have caused serious anxieties in Canberra. Article 10 of the NPT, to which Australia is a signatory, permits a party to withdraw 'if it decides that extraordinary events, related to the subject matter of this Treaty, have jeopardised the supreme interests of its country'. Acquisition of nuclear weapons by Indonesia would have permitted Australia to withdraw from the NPT after a three month notice. Such will not be the case once Australia has ratified the Treaty of Rarotonga. The only grounds for withdrawal from the latter regime is its violation by a treaty member. Since Indonesia is not a party to the SPNFZ, nuclear weapons acquisition by Indonesia will not be a legitimate reason for Australian withdrawal from the SPNFZ regime, whose non-proliferation requirements are more stringent than those of the NPT. Since Australia is the only country with nuclear weapons potential in the South Pacific region, the Treaty of Rarotonga must be credited with at least this substantial achievement in the sphere of non-proliferation.

For all these improvements, the SPNFZ has not escaped criticism. There is a slight apprehension among the conservative tendencies in the region that the zone might be a 'Trojan horse' which will eventually undermine Western security networks. The more substantial attack comes from anti-nuclear groups. Activities of most concern to such groups - port calls, transit facilities for nuclear capable aircraft, command, control and intelligence (C^3I) facilities in Australia and New Zealand, movement of US nuclear forces in the South Pacific waterways, etc. - remain untouched by the treaty. Not only does the treaty carefully avoid impinging upon the substantial US nuclear involvement in the region; its geographical limits specifically exclude US Micronesian territories. Kwajalein Atoll in the Marshall Islands has a permanent US missile testing facility; Guam has a nuclear stockpiling site and a strategic bomber base at Andersen airfield; both have C^3I facilities; and the US has plans for establishing nuclear naval and air bases in Palau and the Mariana Islands.

THE TREATY OF RAROTONGA

Some of these gaps were sought to be plugged during the negotiations over the treaty. PNG and Vanuatu tried unsuccessfully to define stationing in a way that would include limits on the frequency and duration of nuclear ship visits. As it stands at present, the distinction between formal basing and de facto home-porting could be easily blurred; indeed Cockburn Sound in Western Australia hosts a nuclear-armed submarine approximately 25 per cent of the time. Nauru, PNG, the Solomon Islands and Vanuatu tried unsuccessfully to extend Protocol 3 to prohibit testing of nuclear missiles. And Nauru and Vanuatu were equally unsuccessful in seeking to ban uranium exports. In fact in August 1986, Australia decided to resume uranium sales to France, which rather undermines Australia's disarmament credibility in respect of French nuclear testing in the region. Having failed in their efforts to give teeth to the Treaty, Vanuatu and the Solomon Islands led a Melanesian revolt against its softness at the 1986 Forum meeting. Solomon's Prime Minister Sir Peter Kenilorea reiterated shortly before the meeting that until there was a more definite stand against testing and dumping, his government would not be signing the treaty. In effect, and later explicitly, he was calling for the protocols to be signed first by all relevant NWS, including France [23]. Earlier, during a visit to Britain in February, Vanuatu's Prime Minister had accused the Australian government of haste, political expediency, and dishonesty with its own people in suggesting that the whole South Pacific was being denuclearised, when in fact it was not prepared to make even Australia truly nuclear-free [24]. A newly-formed Fiji Anti-Nuclear Group (FANG), campaigning to give more teeth to the SPNFZ as well, held an 'alternative forum' in Suva in August 1986 during the South Pacific Forum meeting; Prime Minister Ratu Sir Kamisese Mara was not amused.

In the light of these 'loopholes' for the entry of nuclear weapons into the South Pacific, the SPNFZ has been described as a 'sham' by various opponents. One peace activist concluded that the SPNFZ Treaty 'seems more a cosmetic measure aimed at containing and defusing growing popular pressure for regional denuclearisation than a serious move towards regional disarmament' [25]. Furthermore, because the loopholes are largely in the categories of growing regional nuclearisation, while the achievements lie in the marginal categories in the South Pacific, there is suspicion that the SPNFZ may represent a net liability in serving to demobilise regional efforts at genuine denuclearisation. As one critic put it, the SPNFZ Treaty 'permits and implicitly legitimises the worst aspects of the nuclear arms race and superpower rivalry in the region' [26]. For these reasons, the Treaty of Rarotonga has been dismissed for having set up a 'Clayton's Zone,' that is, a zone which is not a zone, by joining which countries promise solemnly to avoid doing certain naughty

THE TREATY OF RAROTONGA

things which they had no intention of doing in the first place. It is said to be an agreement by non-nuclear states to stay non-nuclear, akin to a 'Smoke-Free Zone' which applies only to non-smokers. The last comment in particular highlights the unfairness of the criticisms. A 'Smoke-Free Zone' is fully effective when smokers refrain from lighting up within the zone; they are not required to quit smoking permanently, nor even to hand-over all cigarettes and matches to guardians of the zone. Similarly, a NFZ will have achieved its objective if NWS refrain from lighting up their nuclear arsenals within the zone. Elimination of nuclear weapons, while desirable in the long run, is hardly realistic for the moment. This is unlikely to satisfy critics who believe that prohibiting the export of uranium by SPNFZ parties, e.g. Australia, is an eminently realistic non-proliferation goal.

The most serious potential gap in the SPNFZ regime is a lack of machinery for policing compliance with the protocols. A control system is established for verifying the behaviour of treaty parties, but not for monitoring the behaviour of NWS within zonal limits. The experience of the Falklands war is very instructive in this regard [27]. First, Britain used five nuclear-powered submarines as part of its task force during the war: Conqueror, Courageous, Spartan, Splendid and Valiant; the Argentinian cruiser General Belgrano was sunk by the Conqueror on 3 May 1982, with the loss of 368 lives. Article 1 of the Treaty of Tlatelolco restricts the use of nuclear material to peaceful purposes exclusively; Protocol 1 to the Treaty specifically obliges signatories to apply Article 1 of the text to territories under their jurisdiction within the Latin American zone; the Falkland Islands, even if British, are such a territory; Britain has ratified Protocol I. Therefore Britain would appear to have been in breach of its obligations under the Tlatelolco regime. Second, it is far from clear that Britain in fact did not have nuclear weapons aboard some of its vessels in the task force, such as nuclear depth charges. Again there is a prima facie case that Britain violated Protocol 2 by carrying nuclear armaments on the task force. The Falklands war therefore reduced the credibility of NWS assurances by means of signing NWFZ protocols. The Treaty of Rarotonga, in failing to institute procedures for verifying NWS compliance with the three protocols, leaves intact this most important ambiguity in a NFZ regime.

A final weakness in the Treaty of Rarotonga concerns its protocols too. All three protocols are static in that they specify the three and five NWS to whom they are addressed. Should any other country acquire nuclear weapons, their accession would require a formal collective amendment rather than a unilateral signature. The amending procedure is specified in Article 11; a multilateral agreement, even on the apparently most innocuous of statements, can never be taken for granted. This is especially so in light of the fact that any

THE TREATY OF RAROTONGA

amendment to the Treaty of Rarotonga requires a consensus; that is, just one Forum member with close ties to a nuclear power can veto future moves to strengthen the zone.

CONCLUSION

An international development relevant to this article occurred in February 1985, when a two-year UN attempt to increase the number of NWFZs ended in complete collapse. The 21-nation panel, which included all five nuclear powers, was set up to consider the establishment of NWFZs in the Middle East, the Balkans, Northern Europe, Africa, South Asia and the South Pacific. Bhaichand Patel, secretary to the group announced its disbanding because of a failure to reach consensus. He indicated that difficulties arose, not with the major powers, but with countries like India and Argentina, which believed the whole exercise to be an 'unrealistic sideshow' [28]. These two third world leaders were determined to demonstrate that while the NWS did not fulfil Article 6 of the NPT obligations on curbing vertical proliferation, it was not possible to expect non-NWS to reach agreements on limiting horizontal proliferation. The 'real issue' is the weapons stockpile of the nuclear powers; seeking to extend the NPT regime for non-nuclear powers through NWFZs is therefore an attempt to evade the problem rather than solve it by confronting it. The group thus achieved the rare distinction of the Secretary-General of the United Nations failing to receive a report from a body that he had set up.

Regional disarmament can at best supplement universal disarmament. Efforts towards the first are spurred by lack of visible progress in the latter. Because the effects of nuclear war would be global and species-threatening, all countries have both the right and the responsibility to make efforts towards halting and then reversing the nuclear arms race. The hope is that each step such as a NWFZ will lessen the suspicion and distrust that underlies the arms race. The South Pacific zone has already attracted Indonesian attention for possible emulation by Southeast Asia [29]. While comprehensive disarmament remains a long-range goal of the international community, the conviction has grown that immediate and partial measures which would increase confidence and create a more favourable atmosphere for overall disarmament should be pursued. Arms control efforts are open to the criticism that they look too much to the past, that they are reactive and curative. The greater need may well be for measures that are anticipatory and preventive. The SPNFZ can be commended for being prophylactic rather than abortive or therapeutic. This therefore gives it a military significance additional to its political importance as a means of raising the

THE TREATY OF RAROTONGA

Table 4.1: Achievements and Gaps in the SPNFZ

Achievements Prohibited Nuclear Activities	Gaps Permissible Nuclear Activities
1. Nuclear Weapons (NW) acquisition by treaty states	1. NW transit through zone
2. NW (i.e. warheads) testing	2. NW-related C^3I facilities
3. NW basing within treaty states' territorial limits	3. Missile testing
4. NW use/threat against zone	4. Logistic and rest & recreation support for porting and missile testing
5. Dumping of nuclear waste at sea by treaty states	5. Joint exercises with nuclear vessels
	6. Nuclear-armed vessels' treaty
	7. Uranium mining & export
	8. Nuclear power
	9. Nuclear waste disposal on land
	10. Use/threat of NW from within zone by NWS
	11. Stationing & deployment of NW in Micronesia

threshold of nuclear initiation and as a confidence-building measure.

The most important contribution of NWFZs is stabilisation of relations so that the international atmosphere can be relaxed and made more peaceful, and enhancement of confidence that nuclear arms control agreements can be respected and verified. A NWFZ is primarily a means for influencing the nature of peacetime relations between the superpowers. Arguably, global arms control talks are more likely to result from rather than lead to reduction in tensions. Beyond this, governments as well as people must not lose sight of the truth that the course of major wars is littered with broken

THE TREATY OF RAROTONGA

promises. States at war have little time for paper guarantees not backed by force; the world little cares not long remembers treaties violated in times of war. Until such time as a reliable system of international security is in place, therefore, a NWFZ is a useful collateral measure of worldwide arms control. But it is not a policy of national defence. States will be attracted to a NWFZ only when convinced that their vital security interests will be enhanced and not jeopardised by participation. The Treaty of Rarotonga was adopted in 1985 because the South Pacific countries believe that their environmental and national security interests will be promoted by this NFZ regime.

NOTES

1. Text of explanatory memorandum and draft resolution in New Zealand Foreign Affairs Review (NZFAR) 25 (September 1975), pp. 55-7.
2. Ibid. (November 1975), p. 54 and (December 1975), p. 46.
3. Malcolm Templeton, 'Novelty, Gestures and Direction in Foreign Policy: Thoughts from the Twentieth Foreign Policy School', in H. Gold, ed. New Directions in New Zealand Foreign Policy (Auckland: Benton Ross, 1985), p. 142.
4. Australian Foreign Affairs Record (AFAR) 54 (August 1983), pp. 408-9.
5. Test of communique in ibid., 55 (August 1984), pp. 798-802.
6. AFAR 54 (August 1983), p. 409.
7. Ian Brownlie, Principles of Public International Law (Oxford: Clarendon, 1973), pp. 233-5.
8. Otago Daily Times, 9 August 1986.
9. Ibid., 4 September 1986.
10. A. Kuznetsov, 'A Nuclear-Free Zone in the South Pacific and its Importance', International Affairs (Moscow), 1985: 12 (December), pp. 104-7.
11. Canberra Times, 18 March 1986.
12. Otago Daily Times, 27 August 1986.
13. Quoted by Denis Reinhardt, 'Soviet pledge on nuclear-free zone hits US diplomacy', The Bulletin, 1 April 1986, p. 85.
14. Geoff Kitney, 'US gets tough with Australia over nuclear-free zone', National Times, 25 April 1986, p. 5.
15. Otago Daily Times, 28 August 1986.
16. N.Z. Times, 1 September 1985.
17. See Ramesh Thakur, 'A dispute of many colours: France, New Zealand and the "Rainbow Warrior" affair', The World Today (December 1986).
18. 'South Pacific Nuclear-Free Zone'; France: Facts and Figures (Wellington: Embassy of France, June 1985).

19. For an interesting assessment of the long-term health hazards facing Pacific peoples as a result of nuclear pollution, see R.R. Thaman, 'Health and Nutrition in the Pacific Islands'; paper prepared for the Pacific Islands Conference in the Cook Islands, 7-10 August 1985.
20. Otago Daily Times, 30 September 1985.
21. Otago Daily Times, 12 December 1986.
22. National Times (Australia), 30 March 1984, p. 24, quoting leaked departmental documents.
23. Otago Daily Times, 30 July 1986.
24. Canberra Times, 28 February 1986.
25. Michael Hamel-Green, 'South Pacific: A Not-So-Nuclear-Free Zone', Peace Studies (October 1985), p. 6.
26. Ibid., (November/December 1985), p. 42. The article was published in two parts.
27. See J. Gallacher, 'Article VII, The Treaty of Tlatelolco and Colonial Warfare in the 20th Century', Arms Control 5 (December 1984).
28. The Evening Post (Wellington), 11 February 1985.
29. See the comments by Foreign Minister Dr. Mochtar as reported in the Canberra Times, 21 March 1986.

Chapter Five

REGIONAL ARMS CONTROL IN THE SOUTH PACIFIC

Greg Fry

In mid-1985 the South Pacific became the second populated region after Latin America to declare itself a nuclear-free zone. This was done by way of an international treaty which was opened for signature in Rarotonga in the Cook Islands during a meeting of the thirteen member South Pacific Forum. Eight South Pacific leaders signed the treaty before they left Rarotonga; another four promised to sign the document later, leaving Vanuatu's Prime Minister Lini as the only leader who will not be signing the agreement. The treaty was to enter into force when the eighth instrument of ratification had been deposited. This means that what had been just an Australian-sponsored proposal is now a fait accompli.

I propose to consider three issues that have arisen because of the existence of the new treaty. The first concerns the question of the legitimacy of the agreement in the eyes of the international community. In particular, will the nuclear weapons powers give their support to the treaty and does it matter if they do not? The second issue relates to the value of the new treaty as an arms control mechanism. Does it have some substance or should we accept the assertion that it is 'the nuclear-free zone you have when you are not having a nuclear-free zone'? Finally, there is the question of whether a more ambitious regional arms control measure, in the form of a comprehensive nuclear-weapons-free zone should be promoted. To make sense of these issues we first need to review what it is that the treaty seeks to do, and what the treaty represents as a political exercise.

THE SCOPE OF THE TREATY

Let us be clear at the outset that, although the Rarotonga agreement purports to be a 'nuclear-free zone', it is no such thing. It does not establish, or even seek to establish, a zone in which all nuclear activities are prohibited. It is more appropriately described as a partial nuclear-free zone. It is

REGIONAL ARMS CONTROL IN THE SOUTH PACIFIC

primarily an arms control agreement, although it also contains one non-weapons prohibition - a ban on the dumping of radioactive wastes. All other parts of the nuclear fuel cycle are unaffected. The energy, bio-medical and research uses of nuclear technology, for example, are not banned. Even as an arms control agreement the treaty has significant limitations. Some direct weapons involvement and weapons-related activities are unaffected by it. Thus it cannot be said that the Rarotonga treaty even establishes a nuclear-weapons-free zone, let alone the broader concept of a 'nuclear-free zone'.

The cynical interpretation of the Australian Government's use of the 'nuclear-free' label in relation to this treaty is that it saw political capital in giving the impression that the agreement would do more than it in fact does; that is, that it would enhance its disarmament credentials with the peace lobby in the party and the electorate without offending a predominantly pro-ANZUS population. This could well be part of the explanation, but if correct, it represents poor political judgement on the part of the government because use of the 'nuclear-free' label has angered rather than placated the peace lobby within Australia.

While some political opportunity may have been seen in inflating what was a more modest exercise, there is an understandable reason for the use of the misleading 'nuclear-free' title. The Australian Government had initially contemplated the use of the narrower, and more apt, 'nuclear-weapons-free zone'. Although clearly the treaty does not control all nuclear weapons involvement in the South Pacific, it does closely resemble in its broad scope the only precedent for a nuclear-weapons-free zone in a populated area, the Tlatelolco Treaty covering Latin America. Thus a case could be made for calling the treaty a nuclear-weapons-free zone. During the negotiation of the treaty, however, it became clear that the South Pacific states wanted a non-weapons provision included: a ban on the dumping of radioactive wastes. As this meant that the agreement was now to go beyond arms control, the 'nuclear-weapons-free zone' label was dropped in favour of the broader 'nuclear-free zone'.

Specifically, then, what are the arms control objectives of the treaty? Each signatory undertakes:

1. not to manufacture, or otherwise acquire, possess, or have control over, any nuclear explosive device inside or outside the zone, or to seek or receive assistance with such activity, or to give assistance to other states engaged in this activity [1];

2. to prevent the stationing of any nuclear explosive device in its territory, stationing being defined specifically as 'emplantation, emplacement, transportation on land or

REGIONAL ARMS CONTROL IN THE SOUTH PACIFIC

inland waters, stockpiling, storage, installation and deployment [2];

3. to prevent in its territory the testing of any nuclear explosive device and not to assist in the testing activity of any other state [3].

In relation to the latter two undertakings, 'territory' refers to 'internal waters, territorial sea and archipelagic waters, the seabed and subsoil beneath, the land territory and the airspace above them'.

What this adds up to then is a prohibition on the presence of nuclear weapons, or on their manufacture or testing, anywhere within the territories of South Pacific states, up to the twelve mile sea limit. There is one very significant qualification to this general prohibition. The treaty specifically allows each state to make an exception for nuclear weapons that may be aboard ships that are visiting its ports or navigating its territorial sea or archipelagic waters, and for weapons that may be aboard aircraft that are visiting its airfields or which are transiting its airspace [4]. It should be noted that the treaty does not compel signatories to allow such involvement. It leaves the decision to the state concerned.

As can be deduced from what it is that is prohibited, there is no attempt to control nuclear weapons on ships outside the 12 mile territorial limits of South Pacific states or to control weapons on aircraft flying in international airspace. Both are beyond the legal jurisdiction of South Pacific states and are, in any case, activities which are protected by international law. This is enforced by a specific reference in the treaty to the fact that none of its provisions seeks to contravene 'international law with regard to freedom of the seas' [5]. Neither does the treaty seek to control missile testing. The definition of a 'nuclear explosive device' is such that it excludes the delivery system (the missile) if it is not an indivisible part of the weapon. Thus the ban on nuclear weapons testing only refers to explosive devices. The treaty's definition of nuclear weapon also excludes the communication and surveillance facilities which are an integral part of nuclear weapons systems.

In its attempt to ban direct nuclear weapons presence on land while not prohibiting weapons-related activity or the transit of nuclear-armed ships or aircraft, the Rarotonga agreement resembles the Tlatelolco Treaty [6]. The South Pacific treaty goes, however, beyond Tlatelolco in two important respects: it bans so-called 'peaceful nuclear explosions' as well as explosions concerned with weapons testing; and it bans the dumping of radioactive wastes. On the other hand, the Tlatelolco treaty appears to achieve a more complete geographical coverage of its region. This is because nearly all

REGIONAL ARMS CONTROL IN THE SOUTH PACIFIC

the Latin American region is land, which consequently falls within the jurisdiction of zonal states. In the South Pacific most of the region is ocean, therefore falling outside the control of the treaty signatories.

The scope of the treaty in geographical terms is defined by membership of the South Pacific Forum, the regional organisation comprising the independent states of the region. The nuclear prohibitions will therefore almost certainly apply in the following countries: Australia [and its territories], New Zealand [and its territories], Papua New Guinea, the Solomon Islands, Tuvalu, Kiribati, Nauru, Fiji, Western Samoa, Tonga, the Cook Islands and Niue. In addition, there is provision for Britain, France and the United States to sign on behalf of their South Pacific territories. Vanuatu is the only independent country in this region that will not be acceding to the treaty. As its anti-nuclear policies go beyond the provisions of the agreement, it can nevertheless be seen as being in accord with the objectives of the treaty.

Although the actual area of application of the nuclear prohibitions is confined to the territory of the Forum countries, and of the dependencies for which administering powers sign, the treaty defines the geographical scope of the South Pacific nuclear-free zone in a much broader fashion [7]. The boundaries stretch from the border of the Latin American nuclear-weapons-free zone in the east to the west coast of Australia in the west, and from the border of the Antarctic zone in the south to the equator - with some extension into the northern hemisphere to include Kiribati - in the north. This includes a vast area of ocean over which the treaty signatories do not have jurisdiction, and in relation to which the treaty does not seek to apply any nuclear prohibitions. It also includes the French territories which will fall outside the jurisdiction of the treaty unless France signs. This concept of region, then, termed a 'picture frame' approach, represents an intended area of application. It is really a political concept. The fact that French Polynesia is included within it does not mean that nuclear testing will cease there. French Polynesia's inclusion facilitates the political objectives of the treaty members who intend that testing be halted. In addition, the 'picture frame' approach is used to make clear that the South Pacific zone is building onto existing zones in Latin America and Antarctica. The extension of the frame to include high seas over which the treaty has no legal jurisdiction in order that the zone might abut these existing zones is essentially a political exercise.

THE POLITICAL EXERCISE

Like all treaty-making, the negotiation of the Rarotonga agreement was a political exercise. It involved overcoming

REGIONAL ARMS CONTROL IN THE SOUTH PACIFIC

opposition in the region, undertaking a balancing act between the disarmament lobby and the pro-ANZUS electorate in Australia, and, if not pleasing the United States, trying at least not to offend its perception of its essential interests.

The South Pacific nuclear-free zone treaty represents the preferred option of only one country in the region - Australia. To simplify quite complex and often ambiguous positions on this issue, it would seem that the first preference of Vanuatu, New Zealand, the Solomon Islands, and possibly Papua New Guinea would be for a more ambitious arms control arrangement. On the other hand, Fiji, Tonga, Western Samoa, the Cook Islands, and Niue would probably prefer that there be no treaty at all, for fear that it might affect regional security arrangements [8].

This variation in opinion first surfaced at the 1983 South Pacific Forum in Canberra where the new Australian Labour Government first put its proposal before the South Pacific leaders. At the end of this meeting it was clear that it would be a formidable political task to move other South Pacific countries from their preferred positions to one of accepting the middle ground position embodied in the Australian formula [9]. In the event, several developments assisted this process. One important factor was the element of time. South Pacific leaders had only been given short notice of Australia's intention to introduce its proposals at the 1983 Forum, a fact which contributed to their lack of enthusiasm for the concept at that meeting. In the twelve months following the Canberra Forum, however, there was time to explain the provisions and implications of the proposal.

A second development of great importance was the change of government in New Zealand in July 1984. This not only removed one of the influential critics of the scheme, Prime Minister Muldoon; it also introduced a government that was strongly supportive of moves to create a nuclear-weapons-free zone. However, in view of the New Zealand Labour Government's preferred position of establishing a zone in which all nuclear weapons activity was prohibited, this could have proved to be an obstacle to gaining agreement on the less ambitious Australian formula. However, Prime Minister Lange chose the pragmatic course of supporting the Australian proposal, recognising that a more radical initiative would not obtain the same degree of support [10]. New Zealand's influence was not only important in lending general support to the concept; it was also important in making sure that Australia's proposal would be put in treaty form rather than remaining as a 'political concept'. Further, New Zealand lobbied for quick movement towards that goal. On both counts it was successful.

A third critical element in the political exercise of overcoming regional resistance to the Australian formula was the role of Australia's Prime Minister Hawke at the 1984 Forum in

REGIONAL ARMS CONTROL IN THE SOUTH PACIFIC

Tuvalu. The evidence suggests that he brought his considerable skill in consensus-making to bear on the subject, creating a very different outcome to that of the previous Forum. By the end of the meeting, there was unanimous agreement that a draft treaty should be drawn up [11].

From this point the pace was swift. A working group of officials met five times over the following year to flesh out a draft treaty based on the principles and parameters agreed to at the Tuvalu Forum. It was this document that was put before the Prime Ministers at Rarotonga. As the Rarotonga meeting approached, however, there were indications that several countries were having second thoughts about supporting the treaty [12]. It was expected that divisions might occur on the by now familiar lines, with Melanesian countries wanting a more radical zone treaty and Fiji and some of the Polynesian states wanting to exercise caution even in relation to this modest treaty. This, however, did not eventuate. On the day, Vanuatu's Prime Minister Lini was the only leader who felt he could not put his signature on the document.

Although Australia and New Zealand were pushing for a treaty at this particular time and in this particular form, the outcome should not be seen as representing a forceful Australia and New Zealand pushing reluctant Pacific Island countries into signing something that was an anathema to them. There is a longstanding anti-nuclear sentiment throughout the South Pacific Islands. All of these countries actively oppose French nuclear testing and Japan's proposals to dump radioactive wastes in the Pacific and in 1975 they went as far as supporting a New Zealand-inspired proposal for a South Pacific nuclear-free zone [13]. The existence of anti-nuclear sentiment provided a base on which the Australian proposal could build.

The success in obtaining near unanimous agreement was also helped by a number of favourable strategic conditions in the South Pacific. Unlike many other regions, there are no serious tensions between countries or between South Pacific states and countries outside the area, at least none that would prompt a South Pacific state to want to keep open the option in 'go nuclear'. There is also a long record of regional co-operation, and the South Pacific Forum, in particular, provided a useful vehicle for the promotion of such an agreement. Even more important, the region is already nuclear-free in the sense provided for in the treaty, except in the case of the nuclear testing in French Polynesia. This restricted the debate about possible consequences of the treaty in terms of future contingencies. The task was not one of disengaging deployed weapons of superpowers as in Europe, or of dismantling existing bases. All of these factors, together with the strategic isolation of the region, meant that there was at least a basis on which a political exercise could be mounted.

REGIONAL ARMS CONTROL IN THE SOUTH PACIFIC

Fundamentally, the debate within the region turned on the question of whether, and to what extent, American nuclear activities should be allowed. In particular, it focussed on the issue of American nuclear-ship visits to regional ports. The outcome was a treaty that was consciously written so as not to interfere with such activities. It does not place a regional ban on the visits of nuclear-armed ships but leaves the decision whether to do so to national governments. The treaty, then, sits at the edge of the ANZUS debate. To go just one step beyond the formula adopted in the Rarotonga agreement would be regarded as a challenge to ANZUS because it would necessarily affect US involvement in the area. For Australian opposition leader Peacock, even this modest treaty stepped over the mark. For him, it represented 'a further step in the demise of ANZUS' because of its prohibition on homeporting [14].

The outcome represents the highest common factor in regional opinion. It is clear that a proposal seeking to ban any additional nuclear activity just would not have had the support of most of the region. While those preferring no zone could stretch to this moderate proposal they would not have been able to support a zone that went beyond it. For those who wanted a more ambitious zone, the Rarotonga formula represents at least a step in the right direction. It incorporates some of the prohibitions they want to see. It is just that they would go further. The Rarotonga agreement, then, was the only regional arms control treaty that could get support given the current political position of South Pacific countries.

The document also represents a political exercise within Australia. The issue was essentially the same - the degree to which US nuclear involvement should be controlled. The Labour Government came to office with a commitment to ANZUS and to a nuclear-free Pacific [15]. Its subsequent partial nuclear-free zone initiative was the balance between two contradictory objectives. To reflect the majority view in the party and the electorate the Australian proposal had to leave out of the regional initiative any prohibition on United States nuclear activity that would have been seen by Washington or the Australian electorate as constituting the dismantling of the security pact with the United States. A nuclear-free ANZUS was not an option.

ATTITUDE OF THE NUCLEAR WEAPONS POWERS

Now that the treaty has been signed by South Pacific states, the immediate issue is whether it can gain support from beyond the region, and particularly from the five nuclear-weapons powers. They will be asked to sign two protocols attached to the treaty in which they would pledge not to

REGIONAL ARMS CONTROL IN THE SOUTH PACIFIC

contribute to any violation of its provisions, and not to use or threaten to use nuclear weapons against members of the zone. These signatures were not required for the treaty to enter into force. This was to be achieved when eight South Pacific states ratified the agreement. Nor can the nuclear-weapon states legally stop the signatories from prohibiting the nuclear weapons activity outlawed in the treaty. The zonal states have full jurisdiction over the territory to which their undertakings apply.

There are, nevertheless, good reasons for seeking the support of the nuclear-weapons powers. Firstly, there is a need to gain legitimacy for the document in the eyes of the international community. This would assist in achieving one of the objectives of the agreement; namely, to be seen as being of equal status to the Antarctic and Tlatelolco treaties. Secondly, to have the nuclear-weapons powers on side would be preferable to having them work against the zone by influencing member states to break their obligations under the treaty. The promise not to use or threaten to use nuclear weapons against the zone seems less important. It is, after all, only a promise. Once given it could not really make South Pacific states feel any more or less secure.

France, Britain and United States, wearing their other hats as powers administering South Pacific dependencies, are also asked to sign a separate protocol in which they would undertake to extend the treaty prohibitions to these territories. It would be useful if such support were forthcoming in order that all territory in the region be covered by the treaty. British support for this protocol is not so critical given the insignificance of Pitcairn Island; more important would be the inclusion of American Samoa. While French support is not expected in view of its determination to continue nuclear testing in French Polynesia, the act of attempting to obtain its signature on the protocol is nevertheless important. One of the main political objectives of the treaty is to show that France is the only country involved in the area which is not prepared to go along with the antinuclear sentiment of the region, thereby putting further political pressure on France in relation to the testing issue.

The support of the five nuclear weapons powers is, then, desirable but not critical for the treaty's effective operation. Will such support be forthcoming? We should begin with the observation that the South Pacific nuclear-free zone treaty is not the preferred arms control arrangement of any of the nuclear-weapons powers, with the possible exception of China. France is clearly opposed to it. The Soviet Union would like to see a more comprehensive zone banning visits of US ships to regional ports and US communication/surveillance facilities in Australia. The United States and Britain would rather the initiative had not been taken. Only China has indicated support for the development of such a treaty [16].

REGIONAL ARMS CONTROL IN THE SOUTH PACIFIC

This does not mean, however, that these countries will necessarily withhold their support. While the treaty does not represent their preferred option, it may still be the case that they will see their interests to be best served by endorsing the treaty. In fact, there is every possibility that all, except France, will decide to sign the protocols.

The United States has established a set of criteria which it expects a nuclear-free zone initiative to meet before its support is forthcoming. These are:

1. that the initiative should come from the region concerned;
2. that all states in the region should participate;
3. that there should be adequate verification;
4. that existing security arrangements should not be disturbed;
5. that the proposal should prohibit the development or possession of any nuclear explosive device by parties to the treaty; and
6. that the treaty obligations must be consistent with existing international law, and specifically not contravene the principle of freedom of navigation of the high seas and the right of innocent passage through territorial seas [17].

The South Pacific treaty clearly meets criteria 1, 5 and 6. In relation to criterion 2, while Vanuatu and the French territories will not be participating in the treaty, it is unlikely that the US would make this an obstacle unless it was rejecting the treaty for more important reasons. Although the Reagan administration has made the 'adequate verification' question a significant issue in arms control agreements generally, in the South Pacific case the verification procedures should be adequate to police the type of nuclear prohibitions in the treaty. Once again as long as the United States is not looking for an excuse not to sign the treaty, criterion 3 should not provide an obstacle.

The ultimate concerns of the United States are covered in criterion 4. Here the US government will be concerned not just with security arrangements in the region but with the effect on US security interests world-wide. The main question for the United States is whether the treaty is to be seen as an anti-American expression and as a development which could encourage further anti-nuclear sentiment within the region or in other regions under American influence; or whether it legitimises American involvement in the area and will act to contain any move to a more radical nuclear-free zone. Clearly, existing US involvement is not threatened by the treaty. Its concern is more with the symbolism of the initiative, particularly in the wake of the ANZUS crisis induced by New Zealand's anti-nuclear policies.

REGIONAL ARMS CONTROL IN THE SOUTH PACIFIC

Although US Secretary of State Schultz cautioned against the zone before it was signed [18], now that it is a *fait accompli*, the US is more likely to conclude that it is a better tactic to be supportive rather than to draw attention to its dissatisfaction with such a development within its nuclear alliance. The negative effect of the latter response is a lesson that Washington may have learnt from its handling of the New Zealand ships ban issue. Despite British Prime Minister Thatcher's initial negative remarks about the zone, Britain could adopt similar reasoning and therefore be in a position to sign the protocols.

The Soviet Union will be weighing the same issue - whether the treaty is to be seen as one which legitimises US nuclear involvement in the South Pacific or whether it might encourage further developments which would move the region closer to the Soviet Union's preferred option. In the event, the Soviet Union may be influenced by its general policy of supporting nuclear-free zones. Even if it has some concerns about lending legitimacy to a concept that is clearly consistent with continued US nuclear involvement, it may decide that it cannot be seen not to be supporting efforts to create nuclear-free zones. Another question mark hangs over the specific question of whether the Soviet Union will be able to sign a protocol that includes a pledge not to threaten or use nuclear weapons against signatories to the treaty. In view of the strategic significance of the Pine Gap, Nurrungar and Northwest Cape facilities in Australia, the Soviet Union may choose to make a political issue of this undertaking.

The process of obtaining the support of the nuclear-weapons powers, if it is to be judged by the Tlatelolco experience, could take several years [19]. In the meantime we can expect that the rest of the international community will view the new treaty as a legitimate arms control mechanism. It has been designed so as to accord with United Nations guidelines on nuclear-weapons-free zones and is modelled on the only appropriate precedent, the Latin American zone. There should be no difficulty, then, for the treaty to be seen as having equal standing in international law with the Antarctic and Latin American zones. As the only regional arms control treaty since 1967, the Rarotonga agreement should be particularly welcomed by those countries promoting arms control in their own regions: Indonesia and Malaysia in Southeast Asia; Sweden and Finland in Northern Europe; Greece, Yugoslavia and Bulgaria in the Balkans; Pakistan in South Asia; Egypt in the Middle East; and Kenya in Africa. At the same time some countries, such as India and Argentina, which have been critical of the nuclear-free-zone approach to arms control, may be less than enthusiastic about the South Pacific treaty [20].

REGIONAL ARMS CONTROL IN THE SOUTH PACIFIC

VALUE AS AN ARMS CONTROL MECHANISM

From the time the proposal for a nuclear-free zone was first mooted in Australia in mid-1983 it came under attack from the peace movement, on the one hand, and the conservative parties on the other, for being an 'empty concept'. Both Left and Right on the disarmament issue have argued that the zone is a cynical political exercise which does not actually achieve anything in arms control terms. It has been variously described as a 'Clayton's zone' [the zone you have when you are not having a zone], a 'Mickey Mouse zone', a 'joke', a 'farce' and a 'folly' [21]. The starting point for such an assessment is the observation that the list of nuclear prohibitions contained in the treaty does not add up to a nuclear-free zone or even a nuclear-weapons-free zone.

But the argument goes beyond saying that the treaty falls well short of what its 'nuclear-free' title appears to claim for it. It says that the zone does not in fact achieve anything at all in arms control terms. This part of the argument is based on the observation that the treaty will not change any existing weapons involvement because it is left outside the purview of the treaty or, in the case of nuclear testing, the ban will simply be ignored by France. Therefore, it is concluded, there is no arms control value in the concept.

The argument is taken a step further by Senator Valentine representing the Nuclear Disarmament party, and by the Australian Democrats, who claim that the new treaty will actually have negative effects in arms control terms, and that to have no treaty would be better than to have this one [22]. This proposition appears to be based on two premises: firstly, that the zone legitimises American nuclear involvement in the South Pacific, and secondly, the fact that this partial treaty masquerading as a comprehensive nuclear-free zone treaty may stop the push for a more ambitious proposal.

Clearly these critics are correct in pointing out that the treaty falls well short of what is implied in its title, and one can therefore understand why they might see it as a cynical political exercise. But the next stage of the argument is more difficult to accept. The assertion that it has no value at all in arms control terms is based on a false premise: that because it changes nothing, it is therefore an 'empty concept'. It is true to say that it changes no existing involvement but this is to miss a very important objective of arms control. It can have a very important role as a braking mechanism. It does not have to be disarming to be effective.

In the South Pacific case, the preventive role of an arms control mechanism is particularly important because we have an existing situation where there are no nuclear weapons on any territory in the region and no country wants to acquire or develop nuclear weapons. Neither are nuclear forces of the Soviet Union or the United States directly stationed in the

REGIONAL ARMS CONTROL IN THE SOUTH PACIFIC

region. Given that the region is effectively nuclear-weapons-free in this sense, an international agreement in which governments agree to keep it that way - and to enter obligations to that effect backed up by verification procedures - means that an existing favourable situation can be entrenched. This does have value. Similar prohibitions are thought to have value in relation to outer space, Antarctica, the sea-bed and, with some qualifications, Latin America, where related treaties are in force. These treaties do not change any existing nuclear weapons activity, because there is none in any of these environments. The idea is to prohibit weapons involvement before it happens and to base that prohibition on the mutual interests of the nuclear-weapons-powers in containing the geographical spread of their arms competition.

The prohibition on the stationing of foreign nuclear forces in the territory of South Pacific countries is of particular importance in this regard. This extends beyond the obligations that these countries have entered into under the Non-Proliferation Treaty. It is a significant move to effectively ban home-basing in the South Pacific for the nuclear-armed ships or aircraft of either superpower. This puts an obstacle in the way of competitive base development in the South Pacific.

The prohibition on a signatory of the Rarotonga treaty acquiring or manufacturing nuclear weapons itself, may be thought to have no arms control value because all of the signatories of the regional agreement have already entered into such an obligation under the Non Proliferation Treaty. This, however, overlooks the importance of the difference in contexts in which such undertakings are made, and in particular, the difference in the sanctions which might ensure compliance in a regional, as against a global, regime. The regional sanctions could complement the global sanctions to make it less likely that a state breaks out of the non-proliferation regime. Regional sanctions may not only complement global ones; they may also be stronger. A state may be pursuing a number of important political and economic objectives regionally which it would not want to put at risk by abrogating the non-proliferation provisions of a regional treaty entered into with the same states. Thus a regional regime could bring extra sanctions to bear over and above those that might apply at the global level.

In the South Pacific case, this is mainly applicable to Australia as the only potential nuclear-weapons power in the region [23]. It is of some value to have Australia saying to its region as well as to the NPT membership that it will not be acquiring nuclear weapons and is willing to undergo verification of that undertaking. It is also a political signal to countries outside the South Pacific zone which may have to take Australia's actions into account. It constitutes an extra assurance to Indonesia, for example, that Australia is not intending to introduce nuclear weapons into the security equation in the region. Indonesia will know that Australia, having initiated this regime in the South Pacific, could not lightly take a decision that would undercut its own creation. The costs that would be incurred put another obstacle - over and above NPT sanctions - in the way of Australia obtaining nuclear weapons.

If Indonesia, already a signatory to the NPT, made similar undertakings in regard to its region - something it has already indicated it is willing to do by its promotion of a Southeast Asian nuclear-weapons-free zone - then this would be a valuable contribution to the nuclear non-proliferation regime in the area. While Australia and Indonesia would be making their undertakings in relation to two different, though adjacent, regions they would also be signalling to each other their intention to defuse any nuclear competition that could otherwise arise between them. The signals would be backed by the assurance that it would be difficult for each to go against a regional treaty that they had signed unless there were exceptional circumstances that outweighed the diplomatic costs of abrogation. This would seem to be of considerable potential value in constraining a regional nuclear competition. The South Pacific treaty is the first step in this process.

Not only would the South Pacific treaty work well alongside a Southeast Asian treaty, it may be that the existence of the South Pacific treaty will encourage Southeast Asian developments. There are already some signs that this is the case. Indonesia, which along with Malaysia, is promoting the concept in Southeast Asia, has welcomed the South Pacific treaty [24]. ASEAN officials have also been studying the South Pacific initiatives [25].

There are several reasons why the South Pacific zone might encourage these developments. First, it is an adjoining zone. This makes it easier for Indonesia to make such a commitment knowing that Australia has already done so in the South Pacific. Secondly, the Southeast Asian and South Pacific Governments now have extensive contact with each other and the secretariats of their regional organisations now have a limited form of cooperation. Thirdly, the South Pacific treaty formula may suit the Southeast Asian situation - concentrating at first on controlling future superpower base development and constraining proliferation in the region while recognising the political reality of great power presence. Fourthly, the procedures leading to the signing of the document provide an alternative approach to the problem to that adopted in relation to Tlatelolco. Fifthly, there are indications that the Australian Government would like to see the South Pacific treaty influence the formation of a Southeast Asian zone. Australia will thus be encouraging any developments in that direction.

REGIONAL ARMS CONTROL IN THE SOUTH PACIFIC

Developments will not be swift. Despite the statement made by the Malaysian Foreign Minister, Tengku Ahmad Rithaudeen, in September 1984, that an ASEAN senior official's meeting had agreed 'in principle' to establish a nuclear weapons-free-zone [26], the indications are that the proposal is now on the back burner. It has now been put to study by an ASEAN Working Group, not always a sign of progress. The proposal, although firmly supported by Indonesia and Malaysia, has met with an unenthusiastic response from Singapore, Thailand and the Philippines, and there has reportedly been pressure from Washington to drop the idea [27].

Although discussion of a Southeast Asian zone has quietened for the moment, the South Pacific treaty may give it new impetus. At first, the Southeast Asian proposal was more like a comprehensive nuclear weapons free zone, including a ban on nuclear-armed ships passing through Southeast Asian waters, although making certain concessions to the American involvement at Subic Bay and Clark Field. Now, the proposal being talked of is closer to the Tlatelolco model adopted in the Pacific, thus making the Rarotonga formula more useful to ASEAN's purposes. The introduction of a modest arms control regime in Southeast Asia could be an important step both in containing future superpower involvement and in constraining proliferation by states in the region. As the South Pacific treaty may give some encouragement to these developments, this must be counted as part of the potential value of the Rarotonga agreement.

Any assessment of the value of an arms control mechanism must also take account of the verification and compliance provisions. A nuclear-free zone which simply declared the good intentions of participating states to keep an area nuclear-free would have little value. To move beyond a declaration of intent, to an international agreement that will have some chance of actually controlling what it claims to control, requires adequate means of verifying that parties are keeping to their undertaking, and some implicit or explicit sanctions which would tend to ensure compliance, or at least put obstacles in the way of non-compliance.

In the case of the South Pacific treaty, both requirements are met. Verification is provided by the application of IAEA safeguards to peaceful nuclear activities; and by giving a controlling body, called the 'Consultative Committee', the power to direct a special inspection team to investigate any suspected violation on the territory of a member state. The verification process is assisted by a complaints procedure which allows any signatory to the treaty to raise any suspicions of violation with the Consultative Committee [28].

Any attempt by a South Pacific state to acquire nuclear weapons should be picked up by the IAEA inspections. The stationing of foreign nuclear forces would quickly become

REGIONAL ARMS CONTROL IN THE SOUTH PACIFIC

common knowledge, although there could be difficulties in knowing when substantial transiting of ships or planes becomes 'deployment' or 'home-basing'. Nuclear testing could be detected by the seismic monitoring network in Australia and New Zealand. Thus, in terms of what is being prohibited, the verification procedures would seem more than adequate.

The treaty provides no sanctions against non-compliance. As is the case with most arms control agreements, the ultimate sanction is the breakdown of the agreement itself [29]. This is generally regarded as an effective sanction when an arms control agreement is based on the mutual interests of the members in upholding a regime. There is also a broader sanction. To violate this particular treaty is to contribute to the breakdown of the basis of the arms control system generally and even more broadly, the system of international law. That is, even though a state may want to violate a particular treaty, it will have to weigh this action against the effect it will have on the whole system of arms control agreements, the existence of which, in its totality, it will view as being in its interests to uphold. In fact, it is very rare to find cases of clear violation of arms control agreements [30]. States do take their obligations seriously - not through any ethical considerations but because of their interest in supporting the whole system of international law.

There are also other sanctions involved in a regional treaty of this kind. If a state does not comply with its obligations under a regional arms control treaty, this may jeopardise the achievement of other objectives being pursued in the region. In the South Pacific case, there is close cooperation and shared security, economic, and political concerns among the member states. This would put an added obstacle in the way of a state violating its treaty commitments.

A COMPREHENSIVE NUCLEAR-WEAPONS-FREE ZONE?

The principal issue for the future is the question of whether the partial nuclear-weapons-free zone established by the Treaty of Rarotonga could, or should, be replaced by a more ambitious regime. There already exists, throughout the region, a significant, if still minority, view that any arms control regime in the Pacific should be uncompromising in its objectives. This demand is for a zone in which all nuclear weapons and weapons-related involvement, whether on land or sea, is completely prohibited, without exception. As usually stated, it also extends to a ban on all peaceful nuclear activities [31].

The geographical scope of the zone in this comprehensive proposal is also more ambitious than that in the partial nuclear-free zone. It would preferably stretch to Japan and

REGIONAL ARMS CONTROL IN THE SOUTH PACIFIC

Hawaii, but would at least include the United States' Micronesian territories. These Micronesian islands are of particular concern to advocates of a comprehensive nuclear-free zone because of their important involvement in US nuclear weapons systems and because of known contingency plans for future base development. In particular, Kwajalein Atoll in the Marshall Islands provides a permanent missile-testing facility; Andersen Airfield is a strategic bomber base and a storage site for nuclear weapons; and Kwajalein and Guam have weapons-related communication and surveillance facilities. There are also contingency plans to open bases for nuclear-armed ships and aircraft at Palau, and on Tinian and Saipan in the Northern Mariana Islands.

The demand for such a zone has been supported for over a decade by the Nuclear Free and Independent Pacific Movement, a loose coalition of trade unions, churches, women's groups and other non-government interests, from throughout the Pacific Islands [32]. In the 1980s, support for the Movement's objectives has increased dramatically; the growing peace movements in Australia and New Zealand adopted the nuclear-free Pacific demand as a central demand; the New Zealand Labour Party came out strongly for this position, and then in 1984 was able to implement its antinuclear policies in government; the significant Left faction within the Australian Labour Party indicated their support for a radical zone in contrast to the Centre Left and Right positions; and Vanuatu, with strongly anti-nuclear policies, became independent. The position is also supported by influential members of the Papua New Guinea and Solomon Islands Governments.

There are a variety of motives underlying this position: to remove nuclear involvement that might provide a target; to eliminate the threats to the health and safety of Pacific peoples; to constrain superpower rivalry in the area; to cease the region's contribution to nuclear weapons systems which are seen as destabilising, either in the hope of ultimately constraining their development or as simply an ethical position with or without political effect. The effort to promote a nuclear-free Pacific is also closely tied with the issue of the political independence of the region. From the beginning, the Nuclear Free Pacific Movement has seen the nuclearisation of the Pacific as being tied intimately with continuing colonialism. It is seen as no coincidence that the main nuclear involvement in the South Pacific is in the dependent territories of France and the United States.

The call for a nuclear-free Pacific has been influenced significantly by global developments. Like the Peace Movement elsewhere, the South Pacific Movement has been influenced by the destabilising doctrines and technologies promoted by the Reagan Administration; by the geographical spread of weapons on land and sea; by the failure of traditional methods of arms control; by the increased knowledge of the implications of the nuclear infrastructure. They have also been concerned by the increasing military presence in the Pacific.

There are, however, insuperable obstacles in the way of translating the comprehensive nuclear-free zone proposal into an arms control treaty that would actually prohibit all nuclear activities in the area. They fall into two broad areas. The first set of obstacles relate to the fact that nearly all the nuclear activity that a comprehensive zone treaty would seek to prohibit falls outside the legal jurisdiction of the South Pacific States - either in French or American territory, or in international airspace or international waters. This legal fact would be no obstacle if France, the USA, and the Soviet Union, were prepared to agree to extend the prohibition to their territories and forego their right under international law to have their nuclear capable ships and aircraft transit the region. But they are not at present prepared to do this.

Just as France is determined to continue nuclear testing in French Polynesia, so the United States will never close down its Kwajalein missile-testing facility, its strategic base in Guam, or foreclose its contingency options in Tinian, Saipan or Palau, unless the strategic situation changes. If anything, all these operations will become more important, particularly if United States forces have to withdraw from the Philippines. Nor would the United States agree to withdraw its right under international law to have its warships transit the region. And there are no other ways to approach this problem. The United Nations cannot compel these countries to prohibit nuclear activities; nor can a declaration or agreement by countries in the area. This, then, would seem to be an impassable obstacle in the current strategic situation.

This limits the possible coverage of a zone to only those activities over which South Pacific states have jurisdiction. Most of these are already contained in the Rarotonga Agreement. There are five, however, that are not. These are:

1. assistance, by South Pacific countries, with missile testing of a foreign power;
2. port visits by nuclear-armed ships;
3. staging of nuclear-capable aircraft;
4. siting of weapons-related communication/surveillance facilities;
5. the sale of uranium.

The first four of these activities are all to do with American nuclear involvement. The attempt to prohibit them, therefore, would immediately be seen as a challenge to ANZUS, and more broadly, to the region's security connection with the United States. It would be seen as such in Washington, in most of the Pacific Island countries and by the majority of the electorate in Australia. The opposition such a

REGIONAL ARMS CONTROL IN THE SOUTH PACIFIC

move would provoke is formidable. This is not only because of the basically pro-ANZUS view of the region; but also because that view is uncompromising. For ANZUS supporters, any move to constrain United States involvement, and to threaten ANZUS, would be seen as likely to lead to regional insecurity and to make nuclear war more likely by helping to upset the global balance through contributing to a weakening of the West's nuclear alliance. Such a proposal would not, then, gain the support of the Australian electorate or of most South Pacific governments.

There may be political change in the future, however, which will increase support for a proposal that seeks to ban United States involvement. Such developments are possible in the Solomon Islands and Papua New Guinea. We can also expect a growth in such feeling in Australia, spurred on by frustration with the Reagan Administration's weapons policies. But in the short term, it is very unlikely that we would see the fundamental change of thinking in the region that would be involved in a rejection of ANZUS. It would have to be regarded as at least a possible development in the longer term if current US policies continue.

A comprehensive nuclear-weapons-free zone could never be translated from a proposal, or a set of political demands, to a regional arms control treaty that could actually enforce its prohibitions for the reasons enunciated above. It may be, however, that in the longer term, a proposal less ambitious than a comprehensive zone but which banned all nuclear activity within the jurisdiction of the South Pacific states, could get support. Such a move, though, would involve the end of ANZUS and, therefore, a fundamental security reorientation of the countries in this region.

The main development that would move the region in this direction would be the realisation that the nuclear policies of the United States were making nuclear war more, rather than less, imminent. Such a realisation would outweigh the traditional benefits and assumptions underlying the security connection with the United States, because these would become irrelevant in the face of an inevitable global nuclear war. In these circumstances, South Pacific states may decide to withdraw their support for the US nuclear weapons system, a stance which would be symbolic and ethical rather than having much impact on US policies. Thus the key to the future debate on arms control in the South Pacific is the trend in American nuclear policies. The trend under the Reagan Administration to destabilising doctrines and technologies would suggest that there will be increasing support in the South Pacific for a move beyond the Treaty of Rarotonga.

REGIONAL ARMS CONTROL IN THE SOUTH PACIFIC

NOTES

1. South Pacific Nuclear-Free Zone Treaty, Article 3.
2. Ibid., Article 5.
3. Ibid., Article 6.
4. Ibid., Article 5.
5. Ibid., Article 2.
6. For the text of the Tlatelolco Treaty [The Treaty for the Prohibition of Nuclear Weapons in Latin America] see United States Arms Control and Disarmament Agency, Arms Control and Disarmament Agreements, Washington D.C., 1980, pp. 59-81.
7. South Pacific Nuclear-free Zone Treaty, Article 1, and Annex 1.
8. For a more detailed explanation of the positions of Pacific Island countries see G.E. Fry, 'Australia, New Zealand and Arms Control in the Pacific Region', Desmond Ball (ed.), The ANZAC Connection, Sydney: George Allen and Unwin, 1985, pp. 106-108.
9. See, for example, Ian Davis, 'Forum Rejects Nuclear-Free Zone Proposal';, The Age, 31 August 1983, p. 1; C. Brammall, 'Melanesian Alliance Begins To Flex its Political Muscle', Canberra Times, 6 September, 1983, p. 2.
10. 'Lange Allows Nuclear Transit', Dominion, 3 June 1983; transcript of interview between Mr Lange and Mr Tanaka of Asahi Shimbun, 22 August 1984; and Helen Clark, M.P., 'Establishing a Nuclear-free Zone in the South Pacific', Address to the Twentieth Foreign Policy School, Otago University, 18 May 1985, pp. 6-7.
11. Communique of the Fifteenth South Pacific Forum, Funafuti, Tuvalu, 27-28 August 1984, Canberra, Department of Foreign Affairs, News Release, 29 August 1984.
12. See Paul Malone, 'Reservations on N-Free Zone', Canberra Times, 3 August 1985, p. 9. Milton Cockburn, 'Nearly Ready for a Nuclear-free Pacific', Sydney Morning Herald, 7 August 1985, p. 1; and 'Nuclear Ships Row on Boil at ASEAN', Sydney Morning Herald, 11 July 1985, p. 1.
13. 'Sixth South Pacific Forum: Press Communique', Nuku'alofa, Tonga, July 1975, p. 3.
14. Interview with Mr Peacock on the ABC radio programme, 'AM', 8 August 1985. See also 'N-Treaty is Too Risky: Peacock', Sydney Morning Herald, 8 August 1985, p. 1.
15. Australian Labour Party, 1982 Platform Constitution and Rules, p. 81.
16. See Hu Yaobang's address to the National Press Club, Canberra, 16 April 1985, cited in Lee Ngok, 'China and the Strategic Balance in the Asian-Pacific Region', Seminar Paper delivered in the Department of International Relations, Australian National University, August 1985, p. 18.
17. John C. Dorrance, 'Coping With The Soviet Pacific

REGIONAL ARMS CONTROL IN THE SOUTH PACIFIC

Threat', Pacific Defence Reporter (July 1983), p. 27.
 18. Milton Cockburn and Amanda Buckley, 'Pacific Treaty Has US Worried', Sydney Morning Herald, 8 August 1985, p. 1; and Michelle Grattan, 'Hawke Courts Kudos for N-Free Treaty', The Age, 7 August 1985, p. 1.
 19. See John R. Redick, 'The Tlatelolco Regime and Nonproliferation in Latin America', International Organisation, Winter 1981, 35:1, pp. 106-109.
 20. 'Attempt to Expand N-Free Zones Fails', Canberra Times, 11 February, 1985, p. 5.
 21. The 'Clayton's Tonic' Zone was a description first used by Mr Muldoon at the 1983 South Pacific Forum. See also 'Vallentine, Peacock Condemn Agreement', The Australian, 8 August, 1985 p. 5; Howard Conkey, 'Opposition Criticised on Nuclear-free Zone Views', Canberra Times, 16 May 1985, p. 11.
 22. See Georgie Malon, 'Pacific Nuclear Treaty Attacked', Adelaide Advertiser, 8 August 1985, p. 10; and Howard Conkey, 'N-Free Zone Plan "So Defective"' Canberra Times, 29 July 1985, p. 3.
 23. See Desmond Ball, 'Australia and Nuclear Policy', Desmond Ball (ed.) Strategy and Defence: Australian Essays, London. George Allen and Unwin, 1982, pp. 320-327.
 24. See Peter Hastings' interview with the Indonesian Foreign Minister, Dr Mochtar Kusumaatmadja, Sydney Morning Herald, 16 August 1985, p. 11.
 25. James Clad, 'No Nukes Maybe', Far Eastern Economic Review, 28 March 1985, p. 42.
 26. 'ASEAN Declares Nuclear Free Zone', Reuters report, 13 September 1984. This is a report of press statement made by the Malaysian Foreign Minister at the conclusion of the ASEAN Senior Officials' Meeting, Kuala Lumpur, 13 September 1985.
 27. Bruce Dover, 'ASEAN Shelves No-Nukes Policy', Melbourne Herald, 1 April 1985, p. 9; 'ASEAN, ANZUS, PPDA, ZOPPAN or complete NWFZ?' Far Eastern Economic Review, 7 March 1985, pp. 22-23.
 28. South Pacific Nuclear-free Zone Treaty, Articles 8,9.
 29. See Hedley Bull, 'The Problem of Sanctions', in Hedley Bull. The Control of the Arms Race: Disarmament and Arms Control in the Missile Age, New York: Frederick Praeger, 2nd ed. 1965, pp. 215-235.
 30. See Julie Dahlitz, Nuclear Arms Control: With Effective International Agreements, Melbourne: McPhee Gribble, 1983, pp. 18-19.
 31. The most comprehensive statement of this position is contained in Michael Hamel-Green, for the Melbourne Nuclear Free and Independent Pacific Committee, The Case for a Comprehensive Nuclear Free Zone in the South Pacific: A Submission to the South Pacific States, Melbourne, unpublished, 1985.

32. See Rachel Sharp, 'Militarism and Nuclear Issues in the Pacific', Rachel Sharp (ed.), <u>Apocalypse No: An Australian Guide to the Arms Race and Peace Movement</u>, Sydney, Pluto Press, 1984.

Chapter Six

FOR A NUCLEAR-FREE EUROPE

Ken Coates

One day after Mr Francis Pym, the British Secretary of State for Defence, revealed his plans for the placement of 160 cruise missiles in Britain, Russian sources leaked a 'captured' file from American commando headquarters in Europe. This had been stolen almost two decades earlier by a Soviet spy, US Army Sergeant R.L. Johnson. Johnson had later been apprehended by the FBI and sentenced to 25 years in prison. His son subsequently shot him dead during a prison visit. Reporting this rather bizarre story, the Sunday Times (22 June 1980) gave a revealing glimpse, but no more, of what Major-General B.E. Spivy, director of J-3 division, had been organising for the 'defence' of his European allies. Unsurprisingly, perhaps, his schedules included a budget for 'pre-emptive strikes at hundreds of cities in the Soviet Union'. But they also included numerous similar nuclear assaults on 'places in neutral or friendly countries, to deny their resources to Soviet troops'. The lists included named cities: '60 in Yugoslavia, 36 in Austria, 13 in West Germany, 21 in Finland and five in Iran'. A more detailed report was later furnished in the New Statesman (27 June 1980). The documents, it explained, appear:

> to contain over 2,800 targets - possibly double that number - throughout Europe and parts of the Middle East. The targets are not strategic and do not include missile silos, but consist for the most part of lists of airfields and other facilities ... railway and highway bridges, railway marshalling yards and sidings, military headquarters and camps, troop concentrations, waterways, port areas, motorway junctions and major and minor airports ...

The weapons designated for this work ranged from 2.5 kilotons to 1.4 megatons.

All this had been worked out back in 1962. At that time the US Air Force was ready to drop '18 to 20 thousand

FOR A NUCLEAR-FREE EUROPE

megatons of nuclear weapons in Europe and the USSR within a 24-hour period'. Curious observers will note, so prolific was the armoury by the time of the early '60s, that it was thought prudent to assign it to such hitherto unanticipated targets as bridges and motorway junctions.

Up to the mid-1960s the United States had enjoyed a preponderant lead over the USSR in the numbers and refinement of its nuclear weapons. For this reason, during this time, Western military doctrine was based upon 'massive retaliation'. This much was always public knowledge. What was not public was the information that much of this 'retaliation' was designed to forestall any regrettable tendency among allies or neutrals to be over-run. 'Assured destruction' was openly defined by US Defence Secretary Robert McNamara as the capacity to eliminate up to one quarter of the population of the Soviet Union, and up to half its industry. But the proportion of the allied populations scheduled for similar elimination was not public knowledge during the years of those calculations.

Today, we have no way of knowing what secrets are locked in the military planning compounds in either the United States or the USSR. But from what is publicly known it is clear that Europe is in a far worse position at the beginning of the '80s than was even secretly thinkable 25 years ago.

1980 began with an urgent and concerned discussion about rearmament. The Pope, in his New Year Message, caught the predominant mood: 'What can one say', he asked, 'in the face of the gigantic and threatening military arsenals which especially at the close of 1979 have caught the attention of the world and especially of Europe, both East and West?'

War in Afghanistan; American hostages in Teheran, and dramatic pile-ups in the Iranian deserts, as European-based American commandos failed to 'spring' them; wars or threats of war in South East Asia, the Middle East, and Southern Africa: at first sight, all the world in turbulence, excepting only Europe. Yet in spite of itself Europe is at the fixed centre of the arms race; and it is in Europe that many of the most fearsome weapons are deployed. What the Pope was recognising at the opening of the decade was that conflicts in any other zone might easily spill back into the European theatre, where they would then destroy our continent.

Numbers of statesmen have warned about this furious accumulation of weapons during the late '70s. It has been a persistent theme of such eminent neutral spokesmen as the late Olof Palme of Sweden, or President Tito of Yugoslavia. Lord Mountbatten, in his speech, warned that 'the frightening facts about the arms race ... show that we are rushing headlong towards a precipice' [1]. Why has this 'headlong rush' broken out?

First, because of the world-wide division between what is nowadays called 'North' and 'South'. In spite of United

FOR A NUCLEAR-FREE EUROPE

Nations initiatives, proposals for a new economic order which could assist economic development have not only not been implemented, but have been stalemated while conditions have even been aggravated by the oil crisis. Poverty was never morally acceptable, but it is no longer politically tolerable in a world which can speak to itself through transistors, while over and again in many areas, starvation recurs. In others, millions remain on the verge of the merest subsistence. The third world is thus a zone of revolts, revolutions, interventions, and wars.

To avoid or win these, repressive leaders like the former Shah of Iran are willing to spend unheard of wealth on arms, and the arms trade paradoxically often takes the lead over all other exchanges, even in countries where malnutrition is endemic. At the same time, strategic considerations bring into play the superpowers, as 'revolutionary' or 'counter-revolutionary' supporters. This produces some extraordinary alignments and confrontations, such as those between the Ethiopian military, and Somalia and Eritrea, where direct Cuban and Soviet intervention has been a crucial factor, even though the Eritreans have been engaged in one of the longest-running liberation struggles in all Africa: or such as the renewed Indo-China war following the Vietnamese invasion of Cambodia, in which remnants of the former Cambodian communist government appear to have received support from the United States, even though it only came into existence in opposition to American secret bombing, which destroyed the physical livelihood of the country together with its social fabric. A variety of such direct and indirect interventions owes everything to geopolitical expediency, and nothing to the ideals invoked to justify them. Such processes help promote what specialists call the 'horizontal' proliferation of nuclear weapons, to new, formerly non-nuclear states, at the same time that they add their pressure to the 'vertical' proliferation between the superpowers.

Second, the emergence of China into the community of nations (if this phrase can nowadays be used without cynicism) complicates the old pattern of interplay between the blocs. Where yesterday there was a tug-o'war between the USA and the USSR, with each principal mobilising its own team of supporters at its end of the rope, now there is a triangular context, in which both of the old-established contestants may, in future, seek to play the China team. At the moment, the Chinese are most worried about the Russians, which means that the Russians will feel a constant need to augment their military readiness on their 'second' front, while the Americans will seek to match Soviet preparedness overall, making no differentiation between the 'theatres' against which the Russians see a need for defence. It should be noted that the Chinese Government still considers that war is 'inevitable', although it has apparently changed its assessment of the source of the threat [2]. (It is the more interesting, in this context, that the Chinese military budget for 1980 was the only one being substantially reduced, by $1.9 billion, or 8.5 per cent.)

Third, while all these political cauldrons boil, the military-technical processes have their own logic, which is fearsome.

Stacked around the world at the beginning of the decade, there were a minimum of 50,000 nuclear warheads, belonging to the two main powers, whose combined explosive capacity exceeds by one million times the destructive power of the first atomic bomb which was dropped on Hiroshima. The number grows continually. This is 'global overkill'. Yet during the decade, the USA and USSR will have manufactured a further 20,000 warheads, some of unimaginable force.

World military spending, the Brandt Report on North-South economic development estimated, ran in 1980 at something approaching $450 billion a year or around $1.2 billion every day [3]. More recent estimates show that global military expenditures have already passed well beyond this level. Recently both the North Atlantic Treaty Organisation and the Warsaw Treaty Organisation decided to increase their military spending annually over a period of time, by real increments of between 3 per cent and 4.5 per cent each year. That is to say, military outlays are inflation-proofed, so that weapons budgets will automatically swell to meet the depreciation of the currency, and then again to provide an absolute increase. It is primarily for this reason that informed estimates showed that the world-wide arms bill would be more than $600 billion per annum or $1.6 billion each day very early in the 1980s.

As part of this process, new weapons are continually being tested. At least 53 nuclear tests took place in 1979. South Africa also seems to have detonated a nuclear device. New missiles are being developed, in pursuit of the ever more lethal pin-pointing of targets, or of even more final obliterative power. In 1980 the Chinese announced tests of their new intercontinental missile, capable of hitting either Moscow or Los Angeles. The French have released news of their preparations to deploy the so-called 'neutron' or enhanced radiation bomb, development of which had previously been held back by President Carter after a storm of adverse publicity. In the United States, the MX missile, weighing 190,000 pounds and capable of throwing ten highly accurate 350 kiloton (350,000 tons of TNT equivalent) warheads at Russia, each of which will be independently targeted, with high accuracy, is being developed. The R and D costs for this missile in 1981 amounted to $1.5 billion, even before production started. This is more, as Emma Rothschild has complained [4], than the combined research and develop-

FOR A NUCLEAR-FREE EUROPE

ment budgets of the US Departments of Labour, Education and Transportation, taken together with the Environmental Protection Agency, the Federal Drug Administration and the Centre for Disease Control. The MX system was originally designed to run on its own sealed private railway, involving 'the largest construction project in US history' [5]. This plan, if completed, would have 'comprised 200 missiles with 2,000 warheads, powerful and accurate enough to threaten the entire Soviet ICBM force of 1,400 missiles' [6]. No doubt the Russians will think of some suitable response to the MX, at similar or greater expense. As things are, the United States defence budget from 1980-85 amounted to one trillion dollars, and, such is the logic of the arms race, an equivalent weight of new weaponry will have to be mobilised from the other side, if the 'balance' is to be maintained.

All this frenetic activity takes place at a time of severe economic crisis, with many Western economies trapped in a crushing slump and quite unable to expand civilian production. Stagnant or shrinking production provides a poor basis for fierce rearmament, which nowadays often accompanies, indeed necessitates, cuts in social investment, schools, housing and health. The price of putting the Trident system into Britain's arsenal will probably be outbreaks of rickets among our poorer children.

But military research takes priority over everything else, and the result is staggering. In the construction of warheads, finesse now passes any reasonable expectation. A Minuteman III missile carries three multiple independently targetable re-entry vehicles (or MIRVs, as such vehicles are conveniently described) carrying nuclear warheads, and each warhead has an explosive power of 170,000 tons of TNT (170 kilotons, or kt), yet weighs 220 lb. The first atomic bomb ever used in action had an explosive force of 12 kt, and it weighed four tons.

Miniaturisation of megadeath bombs has made fine progress. So has the refinement of delivery systems. This is measured by the standard of Circular Error Probable (CEP), which is the radius of that circle centred on the target, within which it can be expected that 50 per cent of warheads of a given type might fall. Heavy bombers of the Second World War, such as those which visited Hiroshima and Nagasaki, had a very large CEP indeed. The Minuteman III system expects to land half its projectiles within a 350 metre radius of target, having flown more than 8,000 miles to do it. The MX, if it goes according to plan, will have a CEP of only a hundred metres. Such accuracy means that it will be perfectly possible to destroy enemy missile silos, however fortified these might be. The Russians are catching up, however. Their SS18 and SS19 missiles are already claimed to have CEPs of 450 metres.

FOR A NUCLEAR-FREE EUROPE

If rocketry has advanced, so too has experimental aviation. The Americans have already tested Stealth, an aeroplane which 'is virtually invisible to Soviet radar'. Critics say that invisibility has been purchased at the cost of multiple crashes, since the new machines are fashioned into shapes which are decidedly unfunctional for flying, in order to elude detection. Stealth is a fighter, but plans have been leaked (in the course of the American elections, during which, apparently, votes are assumed to be attracted to the most bloodthirsty contender) for a similarly-wrought long-range bomber. Officials in the US Defence Department insist that contorted shapes are only part of the mechanism which defeats radar detection: apparently new materials can be coated onto aircraft skins, to absorb radio waves. By such means, together with navigational advances, it may be hoped to secure even greater accuracy of weapon delivery.

Two questions remain. First, as Lord Zuckerman, the British Government's former chief scientific advisor, percipiently insists, what happens to the other 50 per cent of warheads which fall outside the CEP? The military may not be interested in them, but other people are. Second, this remarkable triumph of technology is all leading to the point where someone has what is politely called a 'first-strike capability'. Both the Russians and the Americans will soon have this capability. But what does it mean? It clearly does not mean that one superpower has the capacity to eliminate the possibility of retaliation by the other, if only it gets its blow in first. What it does signify is the capacity to wreak such destruction as to reduce any possible response to an 'acceptable' level of damage. This is a level which will clearly vary with the degree of megalomania in the respective national leaderships.

All informed commentators are very wary about 'first strike capability' because with it the whole doctrine of mutually assured destruction (appropriately known under the acronym MAD) will no longer apply. With either or both superpowers approaching 'first strike' potential, the calculations are all different. Yesterday we were assured, barring accidents, of safety of a bizarre and frightening kind: but now each new strengthening of the arsenals spells out with a terrifying rigour, a new, unprecedented danger. Pre-emptive war is now a growing possibility. It is therefore quite impossible to argue support for a doctrine of 'deterrence' as if this could follow an unchanging pattern over the decades, irrespective of changes in the political balance in the world, and irrespective of the convolutions of military technology.

In fact, 'deterrence' has already undergone fearsome developments. Those within the great military machines who have understood this have frequently signalled their disquiet. 'If a way out of the political dilemmas we now face is not negotiated', wrote Lord Zuckerman, 'our leaders will quickly

FOR A NUCLEAR-FREE EUROPE

learn that there is no technical road to victory in the nuclear arms race ' [7]. 'Wars cannot be fought with nuclear weapons', said Lord Mountbatten: 'There are powerful voices around the world who still give credence to the old Roman precept - if you desire peace, prepare for war. This is absolute nuclear nonsense' [8].

Yet serious discussion of disarmament has come to an end. The SALT II agreement was not ratified and is now suspended. The Treaty on the non-proliferation of nuclear weapons is breaking down, and the non-nuclear powers are convinced that all the nuclear weapon states are flouting it, by refusing to reduce their nuclear arsenals. It is true that following the initiative of Chancellor Schmidt talks opened between Senator Muskie and Mr Gromyko in order to discover whether negotiations could begin on the reduction of medium range arsenals in Europe. But unless there is a huge mobilisation of public protest, the outcome of such talks is completely predictable.

In spite of detente, and the relatively stable relations between its two main halves during the past decade, Europe remains by far the most militaristic zone of the contemporary world.

At least 10,000, possibly 15,000, warheads are stockpiled in Europe for what is called 'tactical' or 'theatre' use. The Americans have installed something between 7,000 and 10,000 of these, and the Russians between 3,500 and 5,000. The yields of these weapons range, it is believed, between something less than one kiloton and up to three megatons. In terms of Hiroshima bombs, one three megaton warhead would have the force of 250 such weapons. But nowadays this is seen as a 'theatre' armament, usable in a 'limited' nuclear war. 'Strategic' bombs, for use in the final stages of escalation, may be as large as 20 megatons. (Although of course those destined for certain types of targets are a lot smaller. The smallest could be a 'mere 30 or 40 kilotons', or two or three Hiroshimas.) Towns in Europe are not commonly far apart from one another. There exist no vast unpopulated tracts, plains, prairies or tundras, in which to confine a nuclear war. Military installations nestle among and between busy urban centres. As Zuckerman has insisted 'the distances between villages are no greater than the radius of effect of low yield weapons of a few kilotons; between towns and cities, say a megaton'.

General Sir John Hackett, a former commander of the Northern Army Group of NATO, published in 1978 a fictional history of the Third World War [9]. In his book this was scheduled for August 1985, and culminated in the nuclear destruction of Birmingham and Minsk. At this point the Russians obligingly faced a domestic rebellion, and everyone who wasn't already dead lived happily every after. The General, as is often the case, knows a lot about specialised military matters, but very little about the sociology of communism, and not much more about the political sociology of his own side. Of course, rebellions are very likely in every country which faces the immediate prospect of nuclear war, which is why the British Government has detailed contingency plans for the arrest of large numbers of 'subversives' when such a war is about to break out. (These may be discovered, in part, by reference to the secret County War Plans which have been prepared on Government instructions to cope with every problem from water rationing to the burial of the uncountable dead.) But there is no good reason to imagine that subversives are harder to arrest in the USSR than they are in Britain, to put the matter very mildly. Nor is there any very good reason to think that the Soviet Union stands on the brink of revolution, or that such revolution would be facilitated by nuclear war. The contrary may be the case. General Hackett's novel has Poles tearing non-existent communist insignia out of their national flag, and contains a variety of other foibles of the same kind: but we may assume that when it speaks of NATO, it gets things broadly right.

The General discusses the basis of NATO strategy which is known as the 'Triad'. This is a 'combination of conventional defence, battlefield nuclear weapons and strategic nuclear action in closely coupled sequence'. Ruefully, General Hackett continues 'This was as fully endorsed in the United Kingdom as anywhere else in the Alliance. How far it was taken seriously anywhere is open to argument. There is little evidence that it was ever taken seriously in the UK ... an observer of the British Army's deployment, equipment and training could scarcely fail to conclude that, whatever happened, the British did not expect to have to take part in a tactical nuclear battle at all ...' [10].

General Hackett's judgements here are anything but fictional ones. The Earl Mountbatten, in the acutely subversive speech to which we have already referred, spoke of it in the most disparaging terms. If a former Chief of Staff and one-time Chairman of NATO's Military Committee found the idea unbelievable, this is strong evidence that General Hackett is quite right that NATO's basic strategy was indeed not 'taken seriously' in the UK. Yet the doctrine of 'flexible response' binds the UK while it remains in force in NATO, because it is enshrined in NATO's 1975 statement for Ministerial Guidance, in Article 4:

> The long-range defence concept supports agreed NATO strategy by calling for a balanced force structure of interdependent strategic nuclear, theatre nuclear and conventional force capabilities. Each element of this Triad performs a unique role; in combination they provide mutual support and reinforcement. No single element of the Triad can substitute for another. The

FOR A NUCLEAR-FREE EUROPE

concept also calls for the modernisation of both strategic and theatre nuclear capabilities; however, major emphasis is placed on maintaining and improving Alliance conventional forces.

Article 11b develops this beyond any possible ambiguity:

The purpose of the tactical nuclear capability is to enhance the deterrent and defensive effect of NATO's forces against large-scale conventional attack, and to provide a deterrent against the expansion of limited conventional attacks and the possible use of tactical nuclear weapons by the aggressor. Its aim is to convince the aggressor that any form of attack on NATO could result in very serious damage to his own forces, and to emphasise the dangers implicit in the continuance of a conflict by presenting him with the risk that such a situation could escalate beyond his control up to all-out nuclear war. Conversely, this capability should be of such a nature that control of the situation would remain in NATO hands.

Yet so jumpy and jittery are military techniques, and so rapidly does their leapfrog bring both superpowers to the unleashing of ever newer devices, that the settled NATO principles of 1975 were already, in 1979, being qualified:

All the elements of the NATO Triad of strategic, theatre nuclear, and conventional forces are in flux. At the strategic level, with or without SALT, the US is modernising each component of its strategic forces. And, as will be described below, the other two legs of the Triad are being modernised as well.

Integral to the doctrine of flexible response, theatre nuclear forces provide the link between US strategic power and NATO conventional forces - a link that, in the view of many, poses the ultimate deterrent against a European war.

With Strategic parity codified in the recent SALT II agreement, and with major Soviet theatre deployments such as the Backfire bomber and the SS20 missile, some have perceived a loose rung near the top of the flexible response ladder. Thus, consideration is being given to new weapons systems: Pershing II, a nuclear-armed ground-launched cruise missile (GLCM), and a new mobile medium-range ballistic missile (MRBM) [11].

This fateful decision came at the end of a long process of other decisions, beginning with Richard Nixon's arrival in

FOR A NUCLEAR-FREE EUROPE

the United States Presidency. So it was that NATO finally determined, at the end of 1979, upon the installation of nearly 600 new Pershing II and Tomahawk (cruise) missiles. The cruise missiles are low-flying pilotless planes, along the lines of the 'doodlebugs' which were sent against Britain in the last years of Hitler's blitzkrieg, only now refined to the highest degree, with computerised guidance which aspires to considerable accuracy. And, of course, they are each intended to take a nuclear bomb for a distance of 2,000 miles, and to deliver it within a very narrowly determined area. There is a lot of evidence that in fact they don't work in the manner intended, but this will increase no-one's security, because it merely means that they will hit the wrong targets. Some of these might easily be located on 'our' side [12].

President Nixon propounded the doctrine of limited nuclear war in his State of the World message of 1971. The USA, he said, needed to provide itself with 'alternatives appropriate to the nature and level of the provocation ... without necessarily having to resort to mass destruction'.

The conventional notion of deterrence had always been wrapped in swathes of assurances by its proponents that the actual use of nuclear weapons was unthinkable. This had been apparently borne out during the Cuba crisis, of which we have already spoken, when, as one American commentator put it, 'we were eyeball to eyeball with the Russians, and they blinked' [13]. But in today's world, with nuclear forces in the superpowers at near parity, nowadays Time magazine offers up the pious hope that, next time, both parties might blink once. Meantime, so vast are the investments tied into the manufacture of nuclear warheads and their delivery systems that, in any real war, it is not their use but their non-use which has become 'unthinkable'. Since we must still presume that neither major power really wishes to destroy the world, we may begin to understand why more and more weight has therefore been placed on the notion of 'theatre' weapons, which it is canvassed, might be actually employed without annihilating the whole of civilisation.

This is the most urgent of the unlooked-for transformations which have come over the logic of deterrence. It followed the development of highly accurate, adaptable and lethal weapons delivery systems. Now this threatens the very survival of European civilisation. In that crucial last speech of his, to the Stockholm International Peace Research Institute, the Earl Mountbatten seized the heart of the question:

> It was not long, however, before smaller nuclear weapons of various designs were produced and deployed for use in what was assumed to be a tactical or theatre war. The belief was that were hostilities ever to break out in Western Europe, such weapons could be used in

field warfare without triggering an all-out nuclear exchange leading to the final holocaust.

<u>I have never found this idea credible</u> (Coates' emphasis). I have never been able to accept the reasons for the belief that any class of nuclear weapons can be categorised in terms of their tactical or strategic purposes [14].

Lord Zuckerman has also declared that he sees no military reality in what is now referred to as tactical or theatre warfare [15]. 'I do not believe', he told a Pugwash symposium in Canada:

> that nuclear weapons could be used in what is now fashionably called a 'theatre war'. I do not believe that any scenario exists which suggests that nuclear weapons could be used in field warfare between two nuclear states without escalation resulting. I know of several such exercises. They all lead to the opposite conclusion. There is no Marquis of Queensbury who would be holding the ring in a nuclear conflict. I cannot see teams of physicists attached to military staffs who would run to the scene of a nuclear explosion and then back to tell their local commanders that the radiation intensity of a nuclear strike by the other side was such and such, and that therefore the riposte should be only a weapon of equivalent yield. If the zone of lethal or wounding neutron radiation of a so-called neutron bomb would have, say, a radius of half a kilometre, the reply might well be a 'dirty' bomb with the same zone of radiation, but with a much wider area of devastation due to blast and fire [16].

Pressure from the Allies has meant that Presidential statements on the issue of limited war have swung backwards and forwards. At times President Carter gave the impression that he was opposed to the doctrine. But the relevation of 'directive 59' in August 1980 showed that there was in fact a continuous evolution in US military policy, apparently regardless of political hesitations by Governments. Directive 59 was a flat-out regression to the pure Nixon doctrine. As the <u>New York Times</u> put it:

> (Defence Secretary) Brown seems to expand the very meaning of deterrence alarmingly. Typically, advocates of flexible targeting argue that it will deter a sneak attack. But Brown's speech says the new policy is also intended to deter a variety of lesser aggressions ... including conventional military aggression ...

FOR A NUCLEAR-FREE EUROPE

Obviously, as the NYT claims, this is liable to:

increase the likelihood that nuclear weapons will be used [17].

<u>Where</u> would such weapons be used? That place would experience total annihilation, and in oblivion would be unable to consider the nicety of 'tactical' or 'strategic' destruction. If 'limited' nuclear exchanges mean anything at all, the only limitation which is thinkable is their restriction to a particular zone. And that is precisely why politicians in the United States find 'limited' war more tolerable than the other sort, because it leaves a hope that escalation to the total destruction of both superpowers might be a second-stage option to be deferred during the negotiations which could be undertaken while Europe burns. It does not matter whether the strategists are right in their assumptions or not. There are strong reasons why a Russian counter-attack ought (within the lights of the Soviet authorities) to be directed at the USA as well as Europe, if Soviet military strategists are as thoughtful as we may presume. But the very fact that NATO is being programmed to follow this line of action means that Europeans must awaken to understand the sinister change that has taken place, beneath the continuing official chatter about 'deterrence'.

The fact that current Soviet military planning speaks a different language does not in the least imply that Europe can escape this dilemma. If one side prepares for a 'theatre' war in our continent, the other will, if and when necessary, respond, whether or not it accepts the protocol which is proposed for the orderly escalation of annihilation from superpower peripheries to superpower centres. The material reality which will control events is the scope and range of the weapons deployed: and the very existence of tens of thousands of theatre weapons implies, in the event of war, that there will be a 'theatre war'. There may be a 'strategic' war as well, in spite of all plans to the contrary. It will be too late for Europe to know or care.

All those missiles and bombs could never be used in Europe without causing death and destruction on a scale hitherto unprecedented and inconceivable. The continent would become a hecatomb, and in it would be buried, not only tens, hundreds of millions of people, but also the remains of a civilisation. If some Europeans survived, in Swiss shelters or British Government bunkers, they would emerge to a cannibal universe in which every humane instinct had been cauterised. Like the tragedy of Cambodia, only on a scale greatly wider and more profound, the tragedy of post-nuclear Europe would be lived by a mutilated people, prone to the most restrictive and destructive xenophobia, ganging for support into pathetic strong-arm squads in order to club a

FOR A NUCLEAR-FREE EUROPE

survival for themselves out of the skulls of others, and fearful of their own shadows. The worlds which came into being in the Florentine renaissance would have been totally annulled, and not only the monuments would be radioactive. On such deathly foundations, 'communism' may be installed, in the Cambodian manner, or some other more primary anarchies or brutalisms may maintain a hegemony of sorts. What is plain is that any and all survivors of a European theatre war will look upon the days before the holocaust as a golden age, and hope will have become, quite literally, a thing of the past.

This carnivorous prospect is not at all identical with the simple supposition with which supporters of nuclear disarmament are often (wrongly) credited, that 'one day deterrence will not work'. It rather implies that there has been a mutation in the concept of deterrence itself, with grisly consequences for us. In fact, deterrence is now so very costly that 'conventional' responses are becoming impossible, to the point where even superpowers find themselves stalemated unless they are willing to discover means of 'conventionalising' and then actually employing parts of their nuclear arsenals. If the powers want to have a bit of a nuclear war, they will want to have it away from home.

If we do not wish to be their hosts for such a match, then, regardless of whether they are right or wrong in supposing that they can confine it to our 'theatre', we must discover a new initiative which can move us towards disarmament. New technologies will not do this, and nor will introspection and conscience suddenly seize command in both superpowers at once.

We are looking for a political step which can open up new forms of public pressure, and bring into the field of force new moral resources. Partly this is a matter of ending superpower domination of the most important negotiations.

But another part of the response must involve a multinational mobilisation of public opinion. In Europe, this will not begin until people appreciate the exceptional vulnerability of their continent. One prominent statesman, who understood, and drew attention to, this extreme exposure, was Olof Palme. During an important speech at a Helsinki conference of the Socialist International, he issued a strong warning:

> Europe is no special zone where peace can be taken for granted. In actual fact, it is at the centre of the arms race. Granted, the general assumption seems to be that any potential military conflict between the superpowers is going to start some place other than in Europe. But even if that were to be the case, we would have to count on one or the other party - in an effort to gain supremacy - trying to open a front on our continent as well. As Avla Myrdal has recently pointed out, a war can simply be transported here, even though actual

FOR A NUCLEAR-FREE EUROPE

causes for war do not exist. Here there is a ready theatre of war. Here there have been great military forces for a long time. Here there are programmed weapons all ready for action ... [18].

Basing himself on this recognition, Mr Palme recalled various earlier attempts to create, in North and Central Europe, nuclear-free zones, from which, by agreement, all warheads were to be excluded. (We shall look at the history of these proposals below.) He then drew a conclusion of historic significance, which provides the most real, and most hopeful, possibility of generating a truly continental opposition to this continuing arms race:

> Today more than ever there is, in my opinion, every reason to go on working for a nuclear-free zone. <u>The ultimate objective of these efforts should be a nuclear-free Europe</u> (Coates' emphasis). The geographical area closest at hand would naturally be Northern and Central Europe. If these areas could be freed from the nuclear weapons stationed there today, the risk of total annihilation in case of a military conflict would be reduced.

Olof Palme's initiative was launched exactly a month before the 1978 United Nations Special Session on Disarmament, which gave rise to a Final Document that is a strong, if tacit, indictment of the frenetic arms race which has actually accelerated sharply since it was agreed. A World Disarmament Campaign was launched in 1980 by Lord Noel Baker and Lord Brockway, and a comprehensive cross-section of voluntary peace organisations: it had the precise intention of securing the implementation of this Document. But although the goal of the UN Special Session was 'general and complete disarmament', as it should have been, it is commonly not understood that this goal was deliberately coupled with a whole series of intermediate objectives, including Palme's own proposals. Article 33 of the Document reads:

> The establishment of nuclear-weapon-free zones on the basis of agreements or arrangements freely arrived at among the States of the zone concerned, and the full compliance with those agreements or arrangements, thus ensuring that the zones are genuinely free from nuclear weapons, and respect for such zones by nuclear-weapon States, constitute an important disarmament measure [19].

Later, the declaration goes on to spell out this commitment in considerable detail. Article 63 schedules several areas for consideration as nuclear-free zones. They include Africa,

FOR A NUCLEAR-FREE EUROPE

where the Organisation of African Unity has resolved upon 'the denuclearisation of the region', but also the Middle East and South Asia, which are listed alongside South and Central America, whose pioneering treaty offers a possible model for others to follow. Until recently, this was the only populous area to have been covered by an existing agreement, which was concluded in the Treaty of Tlatelolco (a suburb of Mexico City), opened for signature from February 1967.

There are other zones which are covered by more or less similar agreements. Conservationists will be pleased that they include Antarctica, the Moon, outer space, and the seabed. Two snags exist in this respect. One is that the effectiveness of the agreed arrangements is often questioned. The other is that if civilisation is destroyed, the survivors may not be equipped to establish themselves comfortably in safe havens among penguins or deep-sea plants and fish, or live alone upon the Moon.

That is why a Martian might be surprised by the omission of Europe from the queue of continents (Africa, Near Asia, the Far East all in course of pressing; and the South Pacific and Latin America, with the exception of Cuba, already having agreed) to negotiate coverage within nuclear-free zones. If Europe is the most vulnerable region, the prime risk, with a dense concentration of population, the most developed and destructible material heritage to lose, and yet no obvious immediate reasons to go to war, why is there any hesitation at all about making Olof Palme's 'ultimate objective' into an immediate and urgent demand?

If we are agreed that 'it does not matter where the bombs come from', there is another question which is more pertinent. This is: where will they be sent to? Clearly, high-priority targets are all locations from which response might otherwise issue. There is therefore a very strong advantage for all Europe if 'East' and 'West', in terms of the deployment of nuclear arsenals, can literally and rigorously become coterminous with 'USA' and 'USSR'. This would not in one step liquidate the alliances, or end all the tension. But it would constitute a significant pressure on the superpowers, since each would thenceforward have a priority need to target on the silos of the other, and the present logic of 'theatre' thinking would all be reversed. None of this would lift the threat of apocalypse, but it would be a first step in that direction. As things are, we are in the theatre, and they are hoping to be able to watch us burn on their videos.

NUCLEAR-FREE ZONES IN EUROPE

If Europe as a whole has not hitherto raised the issue of its possible denuclearisation, there have been a number of efforts to sanitise smaller regions within the continent.

FOR A NUCLEAR-FREE EUROPE

The idea that groups of nations in particular areas might agree to forgo the manufacture or deployment of nuclear weapons, and to eschew research into their production, was first seriously mooted in the second half of the 1950s. In 1956, the USSR attempted to open discussions on the possible restriction of armaments, under inspection, and the prohibition of nuclear weapons within both German states and some adjacent countries. The proposal was discussed in the Disarmament Sub-Committee of the United Nations, but it got no further. But afterwards the Foreign Secretary of Poland, Adam Rapacki, took to the Twelfth Session of the UN General Assembly a plan to outlaw both the manufacture and the harbouring of nuclear arsenals in all the territories of Poland, Czechoslovakia, the German Democratic Republic and the Federal German Republic. The Czechoslovaks and East Germans quickly endorsed this suggestion.

Rapacki's proposals would have come into force by four separate unilateral decisions of each relevant government. Enforcement would have been supervised by a commission drawn from NATO countries, Warsaw Pact adherents, and non-aligned states. Inspection posts, with a system of ground and air controls, were to be established to enable the commission to function. Subject to this supervision, neither nuclear weapons, nor installations capable of harbouring or servicing them, nor missile sites, would have been permitted in the entire designated area. Nuclear powers were thereupon expected to agree not to use nuclear weapons against the denuclearised zone, and not to deploy their own atomic warheads with any of their conventional forces stationed within it.

The plan was rejected by the NATO powers, on the grounds, first, that it did nothing to secure German reunification and, second, that it failed to cover the deployment of conventional armaments. In 1958, therefore, Rapacki returned with modified proposals. Now he suggested a phased approach. In the beginning, nuclear stockpiles would be frozen at their existing levels within the zone. Later, the removal of these weapon stocks would be accompanied by controlled and mutually agreed reductions in conventional forces. This initiative, too, was rejected.

Meantime, in 1957, Romania proposed a similar project to denuclearise the Balkans. This plan was reiterated in 1968, and again in 1972.

In 1959, the Irish government outlined a plan for the creation of nuclear-free zones throughout the entire planet, which were to be developed region by region. In the same year the Chinese People's Republic suggested that the Pacific Ocean and all Asia be constituted a nuclear-free zone, and in 1960 various African states elaborated similar proposals for an All-African agreement. (These were re-tabled in 1965, and yet again in 1974.)

FOR A NUCLEAR-FREE EUROPE

In 1962 the Polish government offered yet another variation on the Rapacki Plan, which would have maintained its later notion of phasing, but which would now have permitted other European nations to join in if they wished to extend the original designated area. In the first stage, existing levels of nuclear weaponry and rocketry would be frozen, prohibiting the creation of new bases. Then, as in the earlier version, nuclear and conventional armaments would be progressively reduced according to a negotiated timetable. The rejection of this 1962 version was the end of the Rapacki proposals, but they were followed in 1964 by the so-called 'Gomulka' plan, which was designed to affect the same area, but which offered more restricted goals.

Although the main NATO powers displayed no real interest in all these efforts, they did arouse some real concern and sympathy in Scandinavia. As early as October 1961, the Swedish government tabled what became known as the Unden Plan (named after Sweden's Foreign Minister) at the First Committee of the UN General Assembly. This supported the idea of nuclear-free zones and a 'non-atomic club', and advocated their general acceptance. Certain of its proposals, concerning non-proliferation and testing, were adopted by the General Assembly. But the Unden Plan was never realised, because the USA and others maintained at the time that nuclear-free zones were an inappropriate approach to disarmament, which could only be agreed in an comprehensive 'general and complete' decision. Over and again this most desirable end has been invoked to block any less total approach to discovering any practicable means by which it might be achieved.

In 1963, President Kekkonen of Finland called for the re-opening of talks on the Unden Plan. Finland and Sweden were both neutral already, he said, while Denmark and Norway, notwithstanding their membership of NATO, had no nuclear weapons of their own, and deployed none of those belonging to their Alliance. But although this constituted a de facto commitment, it would, he held, be notably reinforced by a deliberate collective decision to confirm it as an enduring joint policy.

The Norwegian premier responded to this demarche by calling for the inclusion of sections of the USSR in the suggested area. As long ago as 1959, Nikita Khrushchev had suggested a Nordic nuclear-free zone, but no approach was apparently made to him during 1963 to discover whether the USSR would be willing to underpin such a project with any concession to the Norwegian viewpoint. However, while this argument was unfolding again in 1963, Khrushchev launched yet another similar proposal, for a nuclear-free Mediterranean.

The fall of Khrushchev took much of the steam out of such diplomatic forays, even though new proposals continue

83

FOR A NUCLEAR-FREE EUROPE

to emerge at intervals. In May 1974, the Indian government detonated what it described as a 'peaceful' nuclear explosion. This provoked renewed proposals for a nuclear-free zone in the Near East, from both Iran and the United Arab Republic, and it revived African concern with the problem. Probably the reverberations of the Indian bang were heard in New Zealand, because that nation offered up a suggestion for a South Pacific free zone later in the year.

Yet, while the European disarmament lobbies were stalemated, the Latin American Treaty, which is briefly discussed above, had already been concluded in 1967, and within a decade it had secured the adherence of 25 states. The last of the main nuclear powers to endorse it was the USSR, which confirmed its general support in 1978. (Cuba withholds endorsement because it reserves its rights pending the evacuation of the Guantanamo base by the United States.) African pressures for a similar agreement are, as we have already argued, notably influenced by the threat of a South African nuclear military capacity, which is an obvious menace to neighbouring Mozambique, Zimbabwe, and Angola, and a standing threat to the Organisation of African Unity. In the Middle East, Israel plays a similar catalysing role, and fear of an Israeli bomb is widespread throughout the region.

Why then, this lag between Europe and the other continents? If the pressure for denuclearised zones began in Europe, and if the need for them, as we have seen, remains direst there, why have the Governments of the Third World been, up to now, so much more effectively vocal on this issue than those of the European continent? Part of the answer surely lies in the prevalence of the non-aligned movement among the countries of the Third World. Apart from a thin scatter of neutrals, Europe is the seed-bed of alignments, and the interests of the blocs as apparently disembodied entities are commonly portrayed as absolute within it. In reality, of course, the blocs are not 'disembodied'. Within them, in military terms, superpowers rule. They control the disposition and development of the two major 'deterrents'. They keep the keys and determine if and when to fire. They displace the constituent patriotisms of the member states with a kind of bloc loyalty, which solidly implies that in each bloc there is a leading state, not only in terms of military supply, but also in terms of the determination of policy. To be sure, each bloc is riven with mounting internal tension. Economic competition divides the West, which enters the latest round of the arms race in a prolonged and, for some, mortifying slump. In the East, divergent interests are not so easily expressed, but they certainly exist, and from time to time become manifest. For all this, subordinate states on either side find it rather difficult to stand off from their protectors.

But stand off we all must. The logic of preparation for a war in our 'theatre' is remorseless, and the profound

FOR A NUCLEAR-FREE EUROPE

worsening of tension between the superpowers at a time of world-wide economic and social crisis all serves to speed up the Gadarene race.

THE EUROPEAN APPEAL

It was in this context that, at the beginning of 1980, the Russell Foundation joined forces with a variety of peace organisations in Britain (CND, Pax Christi, the International Confederation for Disarmament and Peace, among others), to launch an appeal to do precisely this.
An early draft of the appeal was written by E.P. Thompson. This was then circulated very widely in Europe, as a result of which a completely new draft was prepared incorporating the ideas of correspondents in many different countries, in Northern, Western, Southern and Eastern Europe. The appeal then took on its final form after a meeting in London in April 1980 at which French, British, West German and Italian supporters were present.

The appeal was announced at the end of that month and signatures were still being canvassed both in Great Britain and the rest of Europe up to Hiroshima Day, 6 August 1980. It calls upon the two great powers to 'withdraw all nuclear weapons from European territory' and urges the USSR to halt the production of its SS20 missiles at the same time that it calls on the USA not to implement its decision to develop and install Pershing II and cruise missiles in the European 'theatre'. The aim is the removal of all nuclear air and submarine bases, nuclear weapons research, development and manufacturing institutions, and nuclear warheads themselves, from the whole continent, 'from Poland to Portugal'. This convenient slogan does not imply an unwillingness to negotiate the denuclearisation of European Russia, up to the Urals: it merely registers the existing real division between superpowers, East and West, on the one side, and the states of Europe, which are caught up in the effects of the arms race between those powers, like corks bobbing in the flood, on the other.

The initial response to what has already become known as the European Nuclear Disarmament movement was quite extraordinary. Thousands of people signed the launching appeal within a few weeks. It quickly became very evident that European hesitations about nuclear armament were in no way less developed than the reservations of Africans, Latin Americans, or Asians.

The British appeal soon gathered a galaxy of well-known names: more than 60 MPs, a number of representatives of Christian churches and peace movements, several peers, trade union leaders and a remarkable cross-section of men and women from the liberal professions. The composer Peter

FOR A NUCLEAR-FREE EUROPE

Maxwell Davies joined the painter Josef Herman with Moss Evans, the trade unionist. Melvyn Bragg, John Arlott, Glenda Jackson, Susannah York, and Juliet Mills of a new generation joined with earlier anti-nuclear campaigners like J.B. Priestley, Lord Brockway and Peggy Duff. But the growing swell of support included hundreds upon hundreds of other names which were not household words, but which reflected a real movement of opinion. Dockers, coalminers, students, housewives, busmen and computer programmers sent in the printed appeal forms, often affixed to home-made petitions listing dozens of adherents.

The approach to Europe was different. It was not possible and was not desirable for British activists to attempt to prescribe what courses of action were most relevant in all the different nations of the continent. Not only were the individual states within the two blocs quite different, one from another, but each of the important neutral states was caught in its own distinct patterns of affinities, from Yugoslavia, the inspirer of non-alignment, to Spain, recently emancipated from Franco, and poised before a variety of choices about possible future alliances. Facing up to this complexity, the Russell Foundation agreed to a proposal from Arthur Scargill to circulate the British appeal with a separate European call for a conference to elaborate whatever distinct approaches to continental disarmament might be practicable and necessary.

Already, as the British supporters were agreeing on their first moves, mass movements had grown up in Holland, Belgium and Norway. The Dutch and Belgians were able, in the swell of protest, to postpone the implementation of the NATO agreement to install cruise missiles in their countries. An important resistance was developing in Denmark. Very soon after the agreement of the text of the END appeal, contact between these movements took place. Key activists from each of these countries joined with eminent political leaders in supporting the proposed European Convention. In France, the END initiative was announced by Claude Bourdet, the editor of Temoignage Chretienne, at a national press conference in Paris. Similar announcements were made in Oslo and Berlin at the same time, and a variety of organisations and individuals gave their support. From Spain, the well-known euro-communist spokesman Manuel Azcarate was joined by Dr Javier Solana Madariaga of the Socialist Party and by the distinguished Catholic writer Joaquin Ruiz-Gimenez Cortes, together with Joan Miro, the artist. An international meeting was organised by the Spaniards to ventilate the argument in the days before the recalled Helsinki Conference on European Security, scheduled to meet in Madrid in November 1980. Italian support was enthusiastic. An early signatory was Professor Giovanni Favilli, the scientist who had, from the beginning, been actively involved in the

FOR A NUCLEAR-FREE EUROPE

Pugwash movement. He happened to be a councillor in the city of Bologna, and in association with the town's mayor he convened meetings in all of the quartieri (or communal councils) of the region to consider the appeal, which was widely published in the civic journal and elsewhere. A Greek committee began to take shape, with the support of Andreas Papandreou, then opposition leader, amongst others. In Berlin, Professor Ulrich Albrecht, the peace researcher, gave extensive publicity to the appeal and gathered important support for it. At the same time, Rudolf Bahro, the East German author who had been imprisoned for publishing a book in the West and was subsequently deported after an amnesty, brought the adherence of the Green Party. Portuguese supporters included Ernesto Melo Antunes, the former Foreign Minister, and Francisco Marcelo Curto, Minister of Labour in the Soares government, alongside other members of Parliament. By mid-July there were supporters in Finland and Turkey, Hungary and Czechoslovakia, Ireland and Sweden. The roll-call reached from Roy Medvedev in Moscow to Noel Browne, the former Health Minister, in Dublin, and from Gunnar Myrdal in Stockholm to ex-premier Hegedus in Budapest.

The extensive response was reinforced by the encouragement of lateral appeals, in which groups in one country made direct appeals to similar groups in another. Several members of the British TUC General Council wrote to their opposite numbers in France, and a British trade union group soon received approaches from Oslo. Members of British universities appealed to colleagues overseas. As the campaign gathers weight, this kind of initiative will gain importance, because it does not depend upon centralised networks of communications, and is a practical, as well as a symbolic, assertion of a growing all-European consciousness. Efforts were made to 'twin' Cambridge and Siena, which for a time shared the burden of proximity to planned bases for cruise missiles. Such arrangements remain possible at a multiplicity of levels, between women's groups, churches, civic and industrial bodies, students, or a wide variety of sporting and cultural associations.

The problem of East-West contact may be especially susceptible to this kind of treatment. The appeal is quite specific about the need for such meetings:

> we must defend and extend the right of all citizens, East or West, to take part in this common movement and to engage in every kind of exchange.

At the same time, the appeal insists that:

> We must resist any attempt by the statesmen of East or West to manipulate this movement to their own advan-

tage. We offer no advantage to either NATO or the Warsaw alliance. Our objectives must be to free Europe from confrontation, to enforce detente between the United States and the Soviet Union, and, ultimately to dissolve both great power alliances.

No one should seek to minimise the difficulties involved in this task, but neither should they be magnified. On both sides of the European divide, and in the middle of it, there is a very considerable, and rapidly growing, awareness of the great folly of the arms race, which in the appeal is justly styled 'demented'. Whilst there are big differences in the scope for public campaigning in countries which support the two blocs, it is not possible for governments in either camp to ignore the pressure of informed and active public opinion in the other, still less when vast movements are developing a deliberate policy of lateral appeals to one another, and to the peoples of those nations as yet uninvolved in the campaign.

The movement for European nuclear disarmament, then, is a movement to transform the meaning of these blocs, and to reverse their engines away from war. If all previous efforts to denuclearise parts of Europe have foundered, this is in large measure because they have all been partial, and thus incapable of mobilising counter-pressures to dual bloc dominance on a wide enough scale. The paradox is that if President Ceausescu still wants a nuclear-free Balkan Zone, or if President Kekkonen still aspires to a denuclearised Baltic, then both are more likely to succeed within the framework of an all-European campaign than they would be in separate localised inter-state agreements. This is not at all to argue that denuclearisation might not come about piecemeal: of course this is quite possible, even probable. But it will only come about when vast pressures of public opinion have come into being; and these pressures must and will develop, albeit unevenly, over the continent as a whole. Europe is not Antarctica, but Europeans have at least as much right to live as do penguins.

For this reason, people must begin to organise, in a gigantic transnational campaign. If this task is difficult, postponement will not make it easier. A pan-European Convention will only be the beginning, although the organisational problems involved in this first step are large enough in all conscience. There must be meetings of many kinds, demonstrations, cross-frontier marches, concerts and festivals. We must melt the ice which binds our continent. Since, unlike governments, we possess no elaborate lasers or thermo-nuclear devices, we shall have to rely on more democratic powers: the light of our reason, the generosity of our hopes and the warmth of our love for one another. Europe is full of clever, resourceful and kindly people. When they reach out to each other, this daunting labour will shrink to human size, and solving it may then become simplicity itself.

NOTES

Parts of this text have appeared in Smith and Thompson (eds.): Protest and Survive (Penguin, 1981). Other parts appeared in European Nuclear Disarmament (Spokesman Pamphlet No. 72, 1980).

1. Apocalypse Now?, Spokesman, 1980, p. 3.
2. Shortly after this text was first published, the Chinese Communist Party announced a revision of this judgement. War, they now thought, would happen if popular resistance to it proved insufficient to prevent it.
3. Estimates vary markedly, because it is difficult to know what values to assign to Soviet military production costs. If budgets are taken, then Soviet expenditure is apparently greatly reduced, because under a system of central planning prices are regulated to fit social priorities (or cynics might say, Government convenience). The alternative is to cost military output on the basis of world market or United States equivalent prices, which, since the USA still has a much more developed economy than the USSR, would still tend to underestimate the real strain of military provision on the Soviet economy.
4. New York Review of Books, 3 April 1980: 'Boom and Bust', pp. 31-4.
5. Herbert Scoville, Jr: 'America's Greatest Construction: Can it Work?', New York Review of Books, 20 March 1980, pp. 12-17.
6. 'The MX system can only lead to vast uncontrolled arms competition that will undermine the security of the US and increase the dangers of nuclear conflict', said Scoville.
7. Apocalypse Now?, ibid., p. 27.
8. Ibid., p. 13.
9. The Third World War, Sphere Books, 1979.
10. Op.cit., p. 50.
11. NATO Review, No. 5, October 1979, p. 29.
12. The acute problems which this missile has encountered in development makes an alarming story, which is told by Andrew Cockburn in The New Statesman, 22 August 1980.
13. Time, 23 June 1980.
14. Apocalypse Now?, ibid., p. 10.
15. Ibid., p. 21.
16. F. Griffiths and J. C. Polanyi, The Dangers of Nuclear War, University of Toronto Press, 1980, p. 164.
17. Editorial, August 1980.
18. The text of this speech is reproduced in the

FOR A NUCLEAR-FREE EUROPE

Bertrand Russell Peace Foundation's <u>European Nuclear Disarmament - Bulletin of Work in Progress</u>, No. 1, 1980.
19. This document is reproduced in <u>Apocalypse Now?</u>, pp. 41-60.

FOR A NUCLEAR-FREE EUROPE

APPENDIX
EUROPEAN NUCLEAR DISARMAMENT: A MANIFESTO
(April 1980)

We are entering the most dangerous decade in human history. A third world war is not merely possible, but increasingly likely. Economic and social difficulties in advanced industrial countries, crisis, militarism and war in the third world compound the political tensions that fuel a demented arms race. In Europe, the main geographical stage for the East-West confrontation, new generations of ever more deadly weapons are appearing.

For at least 25 years, the forces of both the North Atlantic and the Warsaw alliance have each had sufficient nuclear weapons to annihilate their opponents, and at the same time to endanger the very basis of civilised life. But with each passing year, competition in nuclear armaments has multiplied their numbers, increasing the probability of some devastating accident or miscalculation.

As each side tries to prove its readiness to use nuclear weapons, in order to prevent their use by the other side, new more 'usable' nuclear weapons are designed and the idea of 'limited' nuclear war is made to sound more and more plausible. So much so that this paradoxical process can logically only lead to the actual use of nuclear weapons.

Neither of the major powers is now in any moral position to influence smaller countries to forego the acquisition of nuclear armament. The increasing spread of nuclear reactors, and the growth of the industry that installs them, reinforce the likelihood of world-wide proliferation of nuclear weapons, thereby multiplying the risks of nuclear exchanges.

Over the years, public opinion has pressed for nuclear disarmament and detente between the contending military blocs. This pressure has failed. An increasing proportion of world resources is expended on weapons, even though mutual extermination is already amply guaranteed. This economic burden, in both East and West, contributes to growing social and political strain, setting in motion a vicious circle in which the arms race feeds upon the instability of the world economy and vice versa: a deathly dialectic.

FOR A NUCLEAR-FREE EUROPE

We are now in great danger. Generations have been born beneath the shadow of nuclear war, and have become habituated to the threat. Concern has given way to apathy. Meanwhile, in a world living always under menace, fear extends through both halves of the European continent. The powers of the military and of internal security forces are enlarged, limitations are placed upon free exchanges of ideas and between persons, and civil rights of independent-minded individuals are threatened, in the West as well as the East.

We do not wish to apportion guilt between the political and military leaders of East and West. Guilt lies squarely upon both parties. Both parties have adopted menacing postures and committed aggressive actions in different parts of the world.

The remedy lies in our own hands. We must act together to free the entire territory of Europe, from Poland to Portugal, from nuclear weapons, air and submarine bases, and from all institutions engaged in research into or manufacture of nuclear weapons. We ask the two superpowers to withdraw all nuclear weapons from European territory. In particular, we ask the Soviet Union to halt production of the SS20 medium range missile and we ask the United States not to implement the decision to develop cruise missiles and Pershing II missiles for deployment in Western Europe. We also urge the ratification of the SALT II agreement, as a necessary step towards the renewal of effective negotiations on general and complete disarmament.

At the same time, we must defend and extend the right of all citizens, East or West, to take part in this common movement and to engage in every kind of exchange.

We appeal to our friends in Europe, of every faith and persuasion, to consider urgently the ways in which we can work together for these common objectives. We envisage a Europeanwide campaign, in which every kind of exchange takes place; in which representatives of different nations and opinions confer and co-ordinate their activities; and in which less formal exchanges between universities, churches, women's organisations, trade unions, youth organisations, professional groups and individuals, take place with the object of promoting a common object: to free all of Europe from nuclear weapons.

We must commence to act as if a united, neutral and pacific Europe already exists. We must learn to be loyal, not to 'East' and 'West', but to each other, and we must disregard the prohibitions and limitations imposed by any national state.

It will be the responsibility of the people of each nation to agitate for the expulsion of nuclear weapons and bases from European soil and territorial waters, and to decide upon its own means and strategy, concerning its own territory. These will differ from one country to another, and we do not

FOR A NUCLEAR-FREE EUROPE

suggest that any single strategy should be imposed. But this must be part of a trans-continental movement in which every kind of exchange takes place.

We must resist any attempt by the statesmen of East or West to manipulate this movement to their own advantage. We offer no advantage to either NATO or the Warsaw alliance. Our objectives must be to free Europe from confrontation, to enforce detente between the United States and the Soviet Union, and, ultimately, to dissolve both great power alliances.

In appealing to fellow Europeans, we are not turning our backs on the world. In working for the peace of Europe we are working for the peace of the world. Twice in this century Europe has disgraced its claims to civilisation by engendering world war. This time we must repay our debts to the world by engendering peace.

This appeal will achieve nothing if it is not supported by determined and inventive action, to win more people to support it. We need to mount an irresistible pressure for a Europe free of nuclear weapons.

We do not wish to impose any uniformity on the movement nor to pre-empt the consultations and decisions of those many organisations already exercising their influence for disarmament and peace. But the situation is urgent. The dangers steadily advance. We invite your support for this common objective, and we shall welcome both your help and advice.

Chapter Seven

THE QUEST FOR A BALKAN NUCLEAR-WEAPONS-FREE ZONE

Peri Pamir

In view of the complexities surrounding this issue, the quest for a Balkan Nuclear-Weapon-Free Zone (BNWFZ) has so far proved to be rather elusive. Like any other idea whose feasibility has to be tested in view of existing realities, the first step is to try and ascertain what are the practical possibilities for establishing such a zone in this region ... which is taken to comprise the modern nation states of Albania, Bulgaria, Greece, Romania, Turkey and Yugoslavia. For this, it is necessary to undertake a critical examination of the key issues involved in this debate, proceeding systematically through the following interrelated stages:

1. Identify the preconditions (both general and specific) that would need to be fulfilled in order to create the framework within which a BNWFZ could be realised.

2. Examine whether existing conditions in the region are conducive for the accomplishment of this objective. This would involve an evaluation of both the favourable and the inhibiting elements. More importantly, it would call for an identification of those obstacles that presently hinder progress on this issue.

3. Determine the necessary conditions and the possibilities for securing the removal of the above impeding factors.

4. And finally, on the basis of the above-derived conclusions, estimate the feasibility and prospects for realising the BNWFZ project in the near or medium-term future. This would then bring us to the question of where the situation stands at present, and what would be required in order for endeavours in this direction to eventually bear fruit.

The arguments for and against the BNWFZ proposal are numerous. This essay will not attempt to argue in either direction. Rather it will try and point to certain factors which are highly relevant to any discussion dealing with the realities of the situation. Owing to limitations of time and space, it will confine itself to certain selected aspects of the points stated above without, for example, going into the substance of the matter (i.e., into a detailed examination of existing conditions in the region) as indicated in item 2. Nevertheless, let brief mention be made here of some general attributes of the Balkans region in relation to the plan to set up a NWFZ there.

Broadly speaking, the application of a NWFZ involves an intricate and complex network of several (e.g., military, security, political, social, psychological, etc.) intertwining factors, interacting within and across national boundaries. This generalisation is especially true for the Balkans, a region that has historically been known as the 'tinder box' of Europe by virtue of the instability and conflict that has traditionally reigned there, often staking the competitive interests of the Great Powers, and by the characteristic diversity of its military, political, social, ethnic, religious, cultural and ideological make-up.

Present Balkan realities, which are conditioned as much - if not more - by historic elements and differences as they are by Cold War divisions, attest to the unique character of this region, depicted by the range of positions and perspectives that exist there. Apart from the obvious NATO/WTO split: Yugoslavia is a non-aligned communist state; Albania lives its own exclusive version of Marxist-Leninism in virtual isolation; Bulgaria is a WTO loyalist; Romania and Greece both aspire to greater independence within their own respective defence alliances; while rival NATO members Greece and Turkey have each experienced troublesome relations with their big ally patron the United States, as well as with one another. It is this heterogeneity, deeply imbedded in local tradition, that has made the sporadically sought objective of closer regional cooperation in the political field difficult to realise. This task is further complicated by virtue of existing bilateral disputes (see Sketch on page 96), historic antagonisms, mutual mistrust and the contending strategic interests of both the superpowers - each striving to maintain and/or to enhance its current position and influence - in the region. Consequently, under such conditions of precarious stability where heightened perceptions of security interests are at stake, questions involving multilateral defence arrangements inevitably entail great political uncertainty, caution and risk. In view of these and other constraints, the creation of a NWFZ in the Balkans is generally regarded by most realists as a distant ideal to emerge from two developments: first, from a broader context of closer cooperation and mutual confidence among countries of the region: and second, from the favourable convergence of numerous related and interdependent

BALKAN NUCLEAR-WEAPONS-FREE ZONE

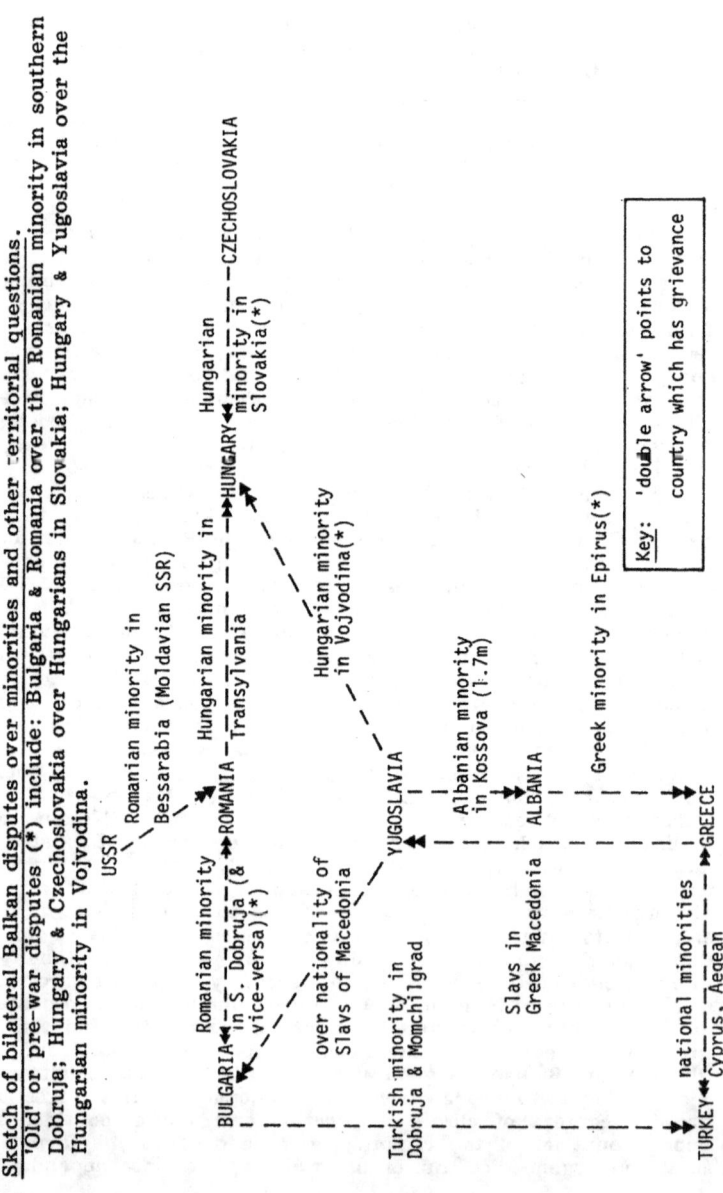

Sketch of bilateral Balkan disputes over minorities and other territorial questions. 'Old' or pre-war disputes (*) include: Bulgaria & Romania over the Romanian minority in southern Dobruja; Hungary & Czechoslovakia over Hungarians in Slovakia; Hungary & Yugoslavia over the Hungarian minority in Vojvodina.

factors which will, in part, be identified in the discussion that follows.

One basic assumption underlying the discussion will be the consideration that in view of the insignificant presence of largely obsolete nuclear weapons systems in the region (see Table 7.1), the setting up of a BNWFZ is essentially a political rather than a military undertaking - although once established, it would have military implications in peacetime. Furthermore, assuming that a BNWFZ agreement would be a form - or a consequence - of multilateral Balkan cooperation, the important question is how to bring about the necessary political collaboration given the wide-ranging diversities that prevail in the region. Consequently, rather than to dwell on the evident virtues of a NWFZ regime for the Balkans, this paper will seek instead, to focus on some factors that now prevent several of the potential partners from making serious advances in the general field of political cooperation - let alone in an area which also involves military/security dimensions. The rationale for this approach is founded on the belief that the overriding objective consists of moving beyond the realm of advocacy and desirability into the pragmatics of trying to chart the possibilities and means for applying the NWFZ concept to the Balkans. The examination of seemingly negative aspects should not be interpreted as an attempt to discredit the BNWFZ proposal. Realistically speaking, it is only once these problems are resolved, or at least recognised, that the chances for realising such a scheme in this area can be properly discerned. More particularly, the identification of existing political constraints constitutes a necessary first step to finding ways whereby some of the more serious obstacles could be removed and the way could be paved for negotiating and implementing an appropriate treaty.

Before dealing with the case of the Balkans, a look at some general attributes and basic preconditions associated with NWFZs should provide the background against which the specifics of the Balkan plan may be drawn.

NWFZs : SOME GENERAL CHARACTERISTICS AND PRECONDITIONS FOR ESTABLISHMENT

Widespread apprehension over the increase in armaments, especially in nuclear arsenals, has compelled the need for some kind of deterrence, equilibrium or balance. The NWFZ idea is an arms control proposal which aims to remove nuclear weapons from specified areas by prohibiting their possession, deployment and use. It consists of an attempt to de-emphasise their importance in military doctrines and defence strategies, as a step towards moderating East-West tension and confrontation. It is believed that in functional terms, a NWFZ would enhance overall stability and regional security by reducing

BALKAN NUCLEAR-WEAPONS-FREE ZONE

Table 7.1: Nuclear-capable Delivery Systems in the Balkan Region

Country	Weapon system	Number	Range (miles)	Yield(KT)	Year of initial development
Greece	F-4 fighter-bomber	56 (est. 1985)	1400(+)	–	1960
	F-104G fighter-bomber	31	1450(+)	–	1968
	155 mm howitzer	240	4–30x10³ metres	2	1972
	8 in. howitzer	n.a.	14–16x10³ metres	1	Early 1950s
	Honest John (surface-to-surface)	8	5–22	20	1951
	Nike-Hercules (surface-to-air)	n.a.	84	1	1958
Turkey	F-4	49 (1978)	As above		
	F-104S	30 (1978)	As above		
	155 mm howitzer	190	As above		
	8 in howitzer	n.a.	As above		
	Honest John	18	As above		
	Nike-Hercules	170	As above		
Romania	Frog SSM	30	10–45	1	Early 1960s
	Scud SSM	20	50	1	1965
Bulgaria	Frog	36	As above		
	Scud	20	As above		
Yugoslavia	155mm howitzer	n.a.	As above		
	Frog-7	n.a.	As above		

n.a.: not available

Source: Platias A.G. & Rydell R.J., "International Security Regimes: The Case of a Balkan Nuclear Weapon-free Zone", in D. Carlton & C. Schaerf, (eds), Arms Control in the 80s, London, Macmillan Press, 1982, Table 10, p. 290.

See page 109 for Note.

BALKAN NUCLEAR-WEAPONS-FREE ZONE

risks of direct involvement in nuclear war and by creating greater confidence. The concept is embodied in a NWFZ framework which presupposes guarantees that the zone is both free of nuclear weapons and immune from the threat of having nuclear weapons used against it. Related to this definition is the implicit understanding that, in principle, only peace-time guarantees can be given [1], the expectation and rationale being that such pledges will help reinforce peace, not remove the risk of nuclear weapons being used against the territory of a non-nuclear-weapon state (NNWS) [2].

Hence, the argument goes that the basic purpose of a NWFZ is not to make this zone a sanctuary in time of nuclear war, but to help prevent such a war from occurring in the first place. In this respect it is recognised that although the zone would have some military significance, its political importance would be far greater by constituting, above all, a major confidence-building measure in the East-West as well as the regional context.

As for those conditions generally associated with the establishment of NWFZs, although in most part interrelated, these are built around the reactions and perceptions of the prospective zonal states on the one hand, and those of the guarantor nuclear-weapon states (NWSs) on the other. Some considerations that would influence potential member states in their decisions about cooperating in forming a NWFZ are as follows.

The most immediate and basic factor affecting local decisions would centre on the perceived security benefits which participation in a NWFZ would be expected to offer. In general terms, for any country considering a NWFZ option the risks associated with such a posture would have to be compensated by the benefits it hopes to derive. For countries already possessing nuclear weapons, this would mean that the perceived disadvantages of possession would have to be seen as overriding its apparent advantages. Under prevailing conceptions of security, however, nations are generally reluctant to renounce those types of armaments with which they fear they may be attacked. Given the widespread tendency to seek conventional and/or nuclear capability as a sign of national will to resist external threats, the choice not only bears on sensitive security interests, but also often presents a dilemma that is far from easy to resolve. The ultimate rationale and motivation for seeking a NWFZ regime would tend to derive from an enhanced security posture, or from the alleviation of existing feelings of insecurity. An individual nation's judgement as to whether a NWFZ would better serve its national security interests would depend on a number of interrelated factors:

1. The extent to which the country is content with its

existing security arrangements would clearly be important.

2. In cases where nuclear weapons are currently deployed, there are different ways in which the country could view this arrangement in terms of its own defence system and strategy. That could be contingent on how great a threat this nation perceived from a NWS, or from a potentially hostile neighbour who had the support of a NWS; that is, it would depend on the importance placed on the 'deterrence' value of current nuclear deployments.

3. In the context of an alliance, an individual country's assessments - as well as its flexibility in affecting 'unilateral' changes in its own defence posture and strategy - would also depend on the role national deployments were considered to play in contributing to the overall balance of forces in the region (and perhaps globally). This would also provide an indication of the amount of resistance a country could be expected to encounter from its allies in the event that it decided to institute changes in its defence posture.

4. The perceptions in items 2 and 3 may in turn be offset by other countervailing factors, including the extent to which a country viewed the increase of nuclear arsenals as a threat to its own security, and/or whether it felt that the potential danger of its becoming a military target in a nuclear conflict would be reduced by adherence to a NWFZ.

5. Related to the above, the absence of a high level of political uncertainty and risk - or, better still, the existence of political acceptability - would need to accompany expected military benefits.

6. Another determining factor would be the degree to which a NWFZ would be compatible with a country's existing alliance commitments, and the manner in which participation in a NWFZ would be likely to affect a nation's relations with its allies and its stature within the alliance. Presumably, a NWFZ should not in any way destabilise existing military alignments or upset relations with allies. In the context of a BNWFZ, this problem is more likely to arise with respect to the NATO countries, for whom it will clearly be more difficult to reconcile participation in a NWFZ with membership in the Western Alliance, given NATO's continued opposition to the concept of NWFZs. Hence, the more general problem would be to find ways to complement membership in the

present two-bloc security system with equally viable and non-conflicting regional alternatives.

7. Apart from the need to obtain the general acquiescence of alliance partners, another crucial condition would be the absence of serious opposition from those major powers which dominated the military situation in the proposed zone area. One fallacy that has at times been associated with the BNWFZ idea is that 'small' or 'medium-sized' allied states could, if necessary, take unilateral policy decisions that defy the expressed wishes of the superpower ally on whom, ultimately, they depend for their security and other related needs. In reality, there are clear limits determining the room for independent action that may be reserved for, or assumed by, smaller allied states in either military bloc. Consequently, the desire for greater independence and flexibility would need to be weighed against the risks and consequences of incurring superpower displeasure. The decision could only be taken by each country on an individual basis in light of existing national choices, realities and priorities.

8. Another related factor affecting national decisions would be the state of the international climate prevailing at the time. In this context, it has been observed that the creation of NWFZs in Europe 'would more likely be a reflection of the existing diplomatic climate than a stimulus for peaceful coexistence' [3]. Consequently, NWFZs are not likely to be 'favoured in time of great strategic uncertainty' [4], or when bloc tensions are considered to be at a high point. Notwithstanding the evident desire on the part of some smaller nations to play a role in promoting peace [5], it is not realistic to think in terms of their making any substantial impact by way of affecting changes in the East-West atmosphere when bloc or superpower tensions remain taut. In cases where these initiatives may not enjoy the full support of their superpower 'patrons', such attempts may even have the adverse effect by creating additional tensions within and across alliances. Given that a BNWFZ cannot be isolated from the broad range of issues related to the East-West situation in Europe, progress on this question will inevitably depend more on the general European political climate than it will on the independent policy initiatives of the Balkan countries themselves.

9. Lastly, it is commonly agreed that non-use - or, negative security - guarantees form an integral part of NWFZ proposals, and that any move in that direction presupposes the consent of all the nuclear weapon states, in

particular the two superpowers, to respect the zone as a first step toward making it a working reality. When dealing in an area such as the Balkans where defence alliances and the superpowers play a predominant role, it must be assumed that undertakings by the major powers in relation to the zone would have important effects on its ultimate success and outcome. Apart from nuclear non-use assurances - including limitations on nuclear deployments in neighbouring areas - it is also argued that, in order to enhance its effectiveness, a NWFZ proposal would have to be accompanied by additional security assurances relating to changes in conventional force structures and military strategies in areas adjacent to the zone. This condition would be particularly relevant to countries that would be renouncing a nuclear-related status under the new arrangement and would be intended to help compensate them for the perception of loss of security associated with such a move.

Coming back to general preconditions, the need to link the withdrawal of nuclear arms to some agreement on changes in conventional force postures is not only necessary to obtaining zonal state adherence but also, and more importantly, to securing the compliance of both superpowers. The superiority generally accorded by NATO to the Warsaw Pact's conventional forces furnishes NATO with continued justification for its strategic doctrine of 'flexible response', with reliance on nuclear retaliation in the event of military conflict in Europe. Thus, under present circumstances any NWFZ proposal would have to be combined with rearrangements of conventional forces in order for the zone to be acceptable to the NATO countries. This may in turn, induce changes in NATO's military doctrine, away from the early use of tactical nuclear weapons. Without such changes, there will be no NWFZs comprising members of the Western Alliance since: 'there is little reason to think that NATO will adopt different nuclear doctrines for different regions of Europe' [6].

Hence, in the context of the NWFZ issue, this debate ultimately boils down to a paradoxical question which underlines the conflicting positions and interests of the two superpowers. On the one hand, could NWFZs, in so far as they would amount to major confidence-building measures, be included among other steps taken independently so as to attain the ultimate goal of nuclear disarmament? Alternatively, could such steps only form part of a more comprehensive arms control agreement between the two blocs? In view of the complexities involved, it is hard to see how the two approaches could be reconciled. At the very least, however, one basic principle that would need to be met if both the super powers are to acquiesce would be for NWFZs not to upset either existing military and security arrangements or the overall region and global balance of strategic forces between the two sides.

An important element related to the incorporation of superpower guarantees in a NWFZ treaty is the need for member states to agree on the role the guarantor powers would be expected to play, or on the extent of their involvement within or in relation to the zone. Closely connected is the need for the zone arrangement to delimit rights and obligations in a clear manner. Ambiguous provisions may provide the guarantor powers with an excuse to exert pressure on member states, or to exercise greater supervision, interference or control which would have the adverse effect of constraining the freedom and independence of the zonal countries. Hence, the danger of lack of clarity in a zone arrangement lies in the possibility of that lack being used as a fine instrument of intimidation. Any vagueness in the 'rights' given to guarantor states could make certain threats, for instance those related to intervention, all the more effective.

The last important general condition for a NWFZ that will be mentioned here is the need for all the major regional powers to agree: (i) to form a zone; and (ii) on the basic objectives the zone is expected to serve. The realisation of common security objectives on a multilateral level presupposes the prior existence of strong and close bilateral ties, and the corresponding absence of mutual mistrust, local tensions or any serious bilateral disagreements among the potential member states.

BNWFZ - OBSTACLES AND PRECONDITIONS

Given this general background, the following is an enumeration of the principal obstacles which hinder prospects for politico-military cooperation in the Balkans and consequently render the realisation of the BNWFZ highly difficult, and for the forseeable future, improbable:

1. Deep-rooted mistrust and the persistence of various bilateral disputes have prevented and most probably will continue to prevent several potential partners from taking serious steps to forge political cooperation. Real progress on wider issues of mutual security - such as a BNWFZ - would therefore appear to be an even lesser likelihood given prevailing hostilities among various regional countries.

In view of what was said earlier about existing constraints on 'small state' room for manoeuvre, a minimum precondition to reducing superpower influence over individual state policies would be for the Balkan countries first to reduce their own differences so as to be able to form an effective 'bloc' which the superpowers would - in some measure - be forced to reckon with. It is difficult to imagine such a development occurring in the foreseeable future when one considers for example, the deep suspicions with long historic roots that divide Yugoslavia and Bulgaria, or Greece and Turkey. As Balkan politics would have it, these two conflicts involve countries within the same broad ideological groupings. Another more recent example would be the controversy that has arisen between Turkey and Bulgaria over the question of Turkish minorities, which has less to do with Cold War divisions than with traditional differences and disaffection arising in part from the predominance accorded to nationalist policies and sentiments on both sides. These disputes - rather like seemingly dead live volcanoes that erupt from time to time - also illustrate the futility of rhetoric which exhorts the need to bury past animosities for the sake of attaining good-neighbourly relations and greater regional cooperation in the future. So long as history dominates perceptions and conditions behaviour, it will take a while before proposals for Balkan 'unity' can realistically 'progress beyond the stage of pious expressions of mutual goodwill' [7].

2. The factor of differing military/political alignments is also likely to present an obstacle given that Balkan countries have traditionally accorded priority to relations with their respective bloc leaders (or other major powers), for reasons of military, political security/dependence, over ties to their neighbours who were similarly weak and often the object of the Balkan countries' own territorial (or other national) ambitions and claims. Consequently, one has to take into account the fact that the recent growth of Balkan 'regional consciousness notwithstanding, national interests and bloc allegiances still take precedence over common political interests'. For this reason, 'any multi-lateral political institutions would tend to reflect the various interests and alignments of the states participating in them and would hardly be adequate guarantees of real security' [8].

This, in turn, raises the question as to what extent allied states are able to take their own interests into consideration when they support or oppose a BNWFZ regime, and to what extent they are also under pressure to abide by the preferences of their superpower allies? The thesis we hold here is that in simple terms, the freedom allied nations have in pursuing preferred policies is very much contingent on the extent to which national and superpower interests converge. This is palpably apparent in the Balkan case where the advantages the WTO countries enjoy by virtue of Soviet support for the proposal contrast sharply with the constraints facing the NATO countries in view of US opposition [9].

3. The nature of varying and often competitive and conflicting national interests and aspirations - which also account for underlying mutual mistrust among some potential member states regarding the true motives and objectives of others in promoting a BNWFZ - presents another main obstacle. Uncertainty and suspicion being mutually reinforcing, this makes each party necessarily wary and distrustful of the other's intentions in this matter. To illustrate the scale of the problem one could, without going into much detail, cite:

Turkish and US mistrust of Soviet intentions
Turkish mistrust of Bulgarian and Greek intentions
Yugoslav mistrust of Soviet and Bulgarian intentions (mutual)
Romanian mistrust of Bulgarian intentions (mutual)
Albanian mistrust of Soviet, Bulgarian and Yugoslav intentions

The prevailing ambiguity over intentions fostered by this vicious circle of mistrust is in fact a good illustration of the framework the BNWFZ proposal is set in. In brief, as long as there is uncertainty regarding the actual purpose a BNWFZ is hoped or expected to serve for each country, the chances for broader political cooperation will be extremely slim.

4. Another constraint centres around prevailing fears that a BNWFZ might introduce changes in the present regional power balance by increasing the prospects for the eventual extension of superpower hegemony over the area. Yugoslavia, for instance, is apprehensive about the effects such an arrangement might have on its precariously balanced posture of non-alignment. especially in the event it led to increased Soviet presence and influence in the region. Albania, forever suspicious of Soviet designs, is similarly sceptical. Turkey also fears it would enhance the position of the USSR and weaken her own defence posture by reducing America's countervailing presence and overall flexibility in the area.

5. A major intractable problem is finding designs that would be acceptable to both the superpowers. As long as the

BALKAN NUCLEAR-WEAPONS-FREE ZONE

US refused to recognise the zone and declined to give guarantees for it, and NATO shied from the responsibility of extending non-nuclear war-time assistance to its allies in the zone, then 'the project is doomed to failure' [10].

In the unlikely event that some countries decide to go ahead despite these constraints (in the belief that a show of determination might induce a reconsideration of original positions), then the actual effectiveness of the zone would be severely compromised by virtue of its failure to cover a geographically contiguous area, and by the continued presence of nuclear weapons on the periphery. These obstacles might eventually be overcome if credible allowances were made to accommodate the constraints facing the NATO countries in the region, and if the initial design contained sufficient in-built flexibility to alleviate the major points of criticism advanced by the Western countries.

Having previously considered certain general preconditions necessary for the successful implementation of NWFZs, an enumeration is offered here of some specific preconditions applicable to the case of the Balkans, taking into account the existing constraints outlined above:

1. An important prerequisite for building a wider system of regional security - and especially one which seeks to restructure defence strategies across alliance boundaries - would be the prior existence of mutual confidence, shared conceptions regarding aims, the political will to collaborate, and the presence of some real sense of harmony of interests among the potential member states. Under present circumstances, the existing rather complex set of bilateral relations among the six relevant countries places limits on efforts to forge closer regional ties. Consequently, any multilateral security pact would need to be preceded by political (and, to a lesser extent, also military) confidence-building efforts designed to dissipate ill-feeling and mistrust resulting from the several bilateral conflicts and disagreements that prevail in the region. Until outstanding bilateral disputes in the area are resolved, it is not possible to envisage inter-Balkan cooperation efforts progressing much beyond their present essentially non-political level.

2. Related to - and following from - the above, in order for regional states to believe that it would be in their mutual interest to cooperate in this manner, they would all need to agree on the nature of the common 'threat' uniting them, and would need to attach equal importance to the 'checking' of this 'threat'. Stated slightly differently, it is important that all parties should essentially be motiv-

ated by the same reason and be in mutual agreement regarding the overriding purpose of the plan.

3. The zone should not conflict with, disrupt or modify alliance memberships and commitments. Likewise, it should not jeopardise individual countries' relations with their major allies.

4. Regional states should not be constrained by opposition from the NWSs, and, in particular, the super-powers.

5. The BNWFZ should not destabilise or threaten the existing balance of forces, whether military or political/-psychological, in the region.

6. To be successful, the zone would have to include all the major actors unless other factors or developments neutralised or overrode this condition.

7. The zone would need to be acceptable to both the super-powers, but a BNWFZ treaty would also need to guard against provisions which might enable the guarantor powers to increase their leverage over the zonal states.

8. Looking at the case of the Balkans, another important condition might well be for the major powers not to get involved in the early stages of regional planning concerning such a sensitive issue. In this context, it would have probably been more prudent for the Soviet Union not to have given its immediate and full support to the last proposal that ostensibly originated in Sofia (in October 1981). Had the zone idea been and remained a purely regional initiative from the beginning, without the imposing shadow of the USSR in the background, it might have had more chance of success. Consequently, one could argue in the light of this observation that one of the conditions considered essential for the successful realisation of a Balkan NWFZ - namely, superpower compliance and support - could also ironically constitute a major barrier if it is not introduced into the picture at the right time or in the right manner.

CONCLUDING REMARKS

This essay has sought to highlight some of the main political perspectives involved in the plan to establish a NWFZ in the Balkans by focussing on the question of preconditions and obstacles. In doing so, it has also tried to provide some indication of when, how, and under what circumstances and conditions such a zone could be set up in this region. One apparent conclusion to be drawn from the preceding analysis

BALKAN NUCLEAR-WEAPONS-FREE ZONE

is that as long as the issues and forces dividing the regional members remain greater and stronger than those uniting them, then the necessary convergence of national aspirations and the required impetus for a collective defence effort will not be generated. What renders this quest particularly elusive is the framework it is set in. That is to say, the question of a BNWFZ is so entrenched in political intangibles that it is extremely difficult, if not impossible under present circumstances, to disengage the technical merits of a zone arrangement from the complex tangle of diverse political factors that presently swamp the problems surrounding this issue. These obstacles which are ingrained into the structure of Balkan politics cannot be ignored. Consequently, any plans concerning denuclearisation of the region will have to be devised in view of rather than in spite of existing realities.

As regards future prospects, in view of recent developments, including the formal rejection by Turkey of the BNWFZ idea [11], prospects for the immediate to medium-term future must be regarded as being rather dim. That is to say, despite widespread concern over the destabilising effects of nuclear escalation and how this might effect security in the Balkans, it is unlikely that these generalised apprehensions will give rise to any significant levels of Balkan regional cooperation in the politico-military sphere. Nevertheless, as long as there is political gain to be had by supporting the concept, the BNWFZ idea is likely be around for some time to come. In the meantime, however, given prevailing interest in furthering existing levels of contact, the most that can be realistically expected in the near-term is an increase in both bilateral and multilateral cooperation in the economic field and other related non-political areas. Initiatives aimed at enhancing good-neighbourly relations through confidence-building measures, including the pursuit of bilateral endeavours to resolve outstanding disagreements should, in turn, create the necessary psychological atmosphere which may eventually pave the way for greater political cooperation in the future.

NOTES

1. As in the case, for example, of the Danish and Norwegian declarations of 1961 prohibiting the stationing of nuclear weapons on their territories in times of peace while simultaneously leaving the war-time option open.
2. In the words of one analyst who draws attention to the basics of the matter, 'nuclear war is unlikely to respect the borders between states that benefit from negative security guarantees and those that do not'. Ultimately, only nuclear disarmament can remove that risk and play a positive and concrete role in enhancing state security. SIPRI Yearbook 1981, World Armaments and Disarmament, Taylor and Francis

Ltd., London, 1981, p. 333.
   3. R.J. Rydell and A. Platias, 'The Balkans : a weapon-free zone?', The Bulletin of Atomic Scientists, May 1982, Vol. 38, No. 5, p. 59.
   4. A.G. Platias and R.J. Rydell, 'International Security Regimes : The Case of a Balkan Nuclear-free Zone', in D. Carlton and C. Schaerf (eds), Arms Control in the 80s, London: Macmillan Press, 1982, p. 295.
   5. A recent example of which would be the 'Declaration of the Six' (i.e., Argentina, Greece, India, Mexico, Sweden and Tanzania), first made on 22 May 1984 and later repeated in January 1985, calling on NWSs to halt all nuclear testing, production and deployment as a 'necessary first step' in promoting disarmament and enhanced security in the world. International Herald Tribune (IHT), 13/8/1984, 29/1/1985.
   6. S. Lodgaard and M. Thee (eds), Nuclear Disengagement in Europe, Taylor and Francis, 1983, p. 55.
   7. R. Clogg, 'Balkan kaleidoscope', The World Today, Vol. 32, August 1976, No. 8, p. 301.
   8. S. Larrabee, 'Balkan Security', Adelphi Papers, No. 135, Summer 1977, p. 37.
   9. To take the case of Bulgaria, for example, despite its earnest support for the denuclearisation plan, it is difficult to imagine the Zhivkov leadership doing anything but complying had Moscow insisted that Sofia accept nuclear missiles as part of the counterdeployments eventually destined for DDR and Czechoslovakia. By the same token, given their security dependence on the US, Greece and Turkey will be confronted with serious dilemmas (as Greece already is) should they decide to cooperate in a BNWFZ project as long as their patron ally remains opposed to the idea.
   10. Lodgaard, op.cit., p. 239.
   11. Milliyet, 4/2/1985 and 18/2/1985; The Economist, 9/1/1985.
   Table 7.1 (p.98). In addition, there are about a dozen or so nuclear storage sites still said to be operating in the two NATO countries. Those in Greece are reportedly not covered by the 1983 US-Greek Defence Agreement and are apparently under exclusive US control (which could be one reason why Papandreou recently agreed to their modernisation for "safety" reasons, having originally refused on grounds that it conflicted with his denuclearisation plans). Similarly, secret annexes were also said to have allowed the US to retain sole control over activities at Greece's Heraklion base. The US-Turkish DCA of 1980 refers to the bases as "joint installations" which doesn't exclude the existence of other secret agreements to cover nuclear facilities strictly under US supervision. Brauch, H.G., 'Confidence-Building Measures in the Balkans and the Eastern Mediterranean', in Carlton, op.cit., p.78; The Economist, 9/2/1985; IHT, 26/3/1985; The New York Times, 23/5/1984.

Chapter Eight

A NUCLEAR-WEAPON-FREE ZONE IN AFRICA?

William Epstein

Every international conference on nuclear energy since the end of World War II has been concerned with two basic goals: first, controlling and eliminating 'atoms for war', and second, promoting and exploiting 'atoms for peace'. The basic dilemma has been that the development of nuclear energy for either purpose also helped to enhance its potential for the other. From the beginning all efforts and negotiations towards the first goal, nuclear-arms control and disarmament, had as their aim not only the control of nuclear weapons but also the prevention of their spread to other countries. The immediate objective was to prevent the further proliferation of nuclear weapons both 'vertically' (i.e., the further development, accumulation and deployment of nuclear weapons by the nuclear powers) and 'horizontally' (i.e., the spread of nuclear-weapon capability to additional powers), and the ultimate objective was to eliminate them entirely.

With the failure of the Baruch Plan in the late 1940s, little hope remained for achieving the elimination of nuclear weapons, but efforts continued to limit their development, to reduce their numbers and above all to prevent their horizontal proliferation.

Two different approaches to the problem of preventing the spread of nuclear weapons were developed: first, the creation of nuclear-weapon-free zones (NWFZs) in which all nuclear weapons would be prohibited, and second, the enactment of a treaty that would specifically ban the dissemination of nuclear weapons by the nuclear powers and the acquisition of such weapons by states not possessing them. These two approaches were supplementary to the ban on nuclear-weapons tests, which was considered as being an important aim in

---

In this essay the expression 'nuclear powers', 'non-nuclear powers', are used interchangeably with the more cumbersome 'nuclear-weapon powers', 'non-nuclear-weapon powers'.

A NUCLEAR-WEAPON-FREE ZONE IN AFRICA

itself but which it was recognised would also help to prevent the proliferation of nuclear weapons.

After 1958, when Poland first proposed the Rapacki plan for the denuclearisation of Central Europe (Poland, Czechoslovakia, East Germany and West Germany), various proposals were put forward for the denuclearisation of other geographic areas, including the Balkans (by Romania), the Mediterranean (by the Soviet Union), the Middle East (by Iran), the Nordic countries (by Finland) and Asia and the Pacific region (by China). These proposals consisted mainly of general concepts rather than concrete steps. Formal and specific plans were put forward only for Central Europe, Africa and Latin America. All these early proposals, except the one concerning Latin America, failed to make significant progress because of the complex political and strategic questions involved.

More recently, after the Indian explosion of a so-called 'peaceful' nuclear device on May 18, 1974, the proposal for a NWFZ in the Middle East was revived by Iran and Egypt, and that for a NWFZ in Africa was revived by Nigeria together with a number of other African states. In addition, new proposals were put forward for a NWFZ in South Asia by Pakistan, and for a zone in the South Pacific by Fiji, New Zealand and Papua-New Guinea. In the latter case, a South Pacific nuclear-free zone was actually declared in 1985. The idea of NWFZs has become politically fashionable.

Almost all proposals for NWFZs have been supported in principle by the Soviet Union and its allies, but they were mainly interested in Central Europe, the Mediterranean and Asia, where the two great-power blocs confronted each other and where the danger of nuclear conflict seemed greatest. Proposals for NWFZs in those three regions were initiated by the Soviet Union or its allies and were aimed mainly at reducing the US nuclear presence there. The United States and its allies considered that such reduction of US military power would give some military or political advantage to the Soviet Union. Accordingly, they conceived of such zones chiefly in the context of preventing the spread of nuclear weapons, and they laid down certain principles, regarding their creation:

1. that they should not upset the existing military balance;
2. that they should be initiated by the states in the region;
3. that they should include all the countries of the area if possible, or at least those with significant military power; and
4. that they should be subject to verification to ensure that the zone would remain nuclear-free.

A NUCLEAR-WEAPON-FREE ZONE IN AFRICA

THE LATIN AMERICAN NUCLEAR-WEAPON-FREE ZONE

Until recently, the only NWFZ created specifically for the purpose of preventing the spread of nuclear weapons was the one established by the Treaty for the Prohibition of Nuclear Weapons in Latin America (commonly referred to as the Treaty of Tlatelolco, after the borough of Mexico City where it was signed on February 14, 1967). Other NWFZs - those established by the Antarctic Treaty of 1959, the Outer Space Treaty of 1967 and the Seabed Treaty of 1971 - were created in part for the purpose of arms limitation, but they were mainly concerned with the efforts of the world community to regulate the use of these still unexplored, unexploited and uninhabited environments. Unlike the Treaty of Tlatelolco, their arms limitation aspects were secondary and therefore comparatively easy to achieve.

The parties to the Treaty of Tlatelolco undertook to prohibit and prevent in their territories the testing, use, production, acquisition, storage, deployment and any form of possession, directly or indirectly, of any nuclear weapons. Protocol I provided for the foreign powers that were responsible for territories within the zone to undertake to apply the Treaty in those territories. Protocol II provided for the nuclear weapon states to undertake to abide by and observe the nuclear free status of the zone and not to use or threaten to use nuclear weapons against the parties to the Treaty.

As regards the control provisions of the Treaty, a comprehensive system of verification was provided, which included the application of the International Atomic Energy Agency (IAEA) safeguards system, periodic reports of the parties to the agency established to implement the treaty, Organisation for the Prohibition of Nuclear Weapons in Latin America (OPANAL), special reports when requested by the General Secretary of OPANAL, and special inspections in addition to the IAEA safeguards system in the case of suspicion of violation.

It was also agreed that nuclear materials and facilities would be used exclusively for peaceful purposes, and that nuclear explosions for peaceful purposes could be carried out under international observation, including explosions involving a device similar to that used in nuclear weapons, so long as such device was not equivalent to a nuclear weapon. The Treaty of Tlatelolco, unlike the 1963 Partial Test Ban Treaty and the Non-Proliferation Treaty (NPT), does provide a definition of nuclear weapons. Article 5 of the Treaty provided that a 'nuclear weapon is any device which is capable of releasing nuclear energy in an uncontrolled manner and which has a group of characteristics that are appropriate for use for warlike purposes'. There is a difference of opinion as to whether the Treaty thus, in effect, bars peaceful nuclear

A NUCLEAR-WEAPON-FREE ZONE IN AFRICA

explosions, but the majority view is that, under current technology, such explosions are banned.

The Treaty, unlike the NPT, contains no explicit provision for the promotion of the peaceful uses of nuclear energy, but Article 17 provides that 'nothing ... in this Treaty shall prejudice the rights of the contracting parties ... to use nuclear energy for peaceful purposes...'

Twenty-six states of Latin America and the Caribbean have now signed the Treaty and 25 of them have ratified it. Of those 25 states the Treaty is in force for 23, each of which has deposited a declaration of waiver of the rather far-reaching requirements for entry into force. Although Brazil and Chile have ratified the Treaty, they have not deposited the declaration of waiver, and the Treaty is therefore not in force for them. It is also not in force for Argentina, which has signed the Treaty but has not ratified it. Cuba has neither signed nor ratified the agreement. The Netherlands, Great Britain and the United States have signed and ratified Protocol I to the treaty, so that it applies to their territories in the zone. France has signed but has not yet ratified Protocol I. Both the United States and France were slow to sign this protocol, prompting several resolutions from the UN General Assembly.

Protocol II has been signed and ratified by China, France, Great Britain, the United States and the Soviet Union, so that they are bound to respect the zone and not use or threaten to use nuclear weapons against it.

The Soviet Union was slow to sign Protocol II, in spite of a number of resolutions by the General Assembly calling on it to do so. Among the reasons given is that the Treaty allows peaceful nuclear explosions, that it does not apply to the Panama Canal Zone or to the transit of nuclear weapons and that the area of the zone takes in too large a part of the Atlantic and the Pacific. The Soviet Union said, however, that it would respect the nuclear-free status of each state in the area that remains nuclear-free.

At the time, The Treaty of Tlatelolco established the only NWFZ in any populated area of the world. The General Assembly hailed it as of historic importance and as a possible precedent for other regions. It encompasses more than 7.5 million square miles inhabited by some 200 million people. What is more important, the number of its parties and supporters, including the nuclear signatories of the Protocols, keeps growing year by year. Such potential nuclear powers as Mexico, Venezuela, Colombia and Peru are full parties. It is noteworthy that Colombia is not a party to the NPT.

If there is any chance at all of Argentina and Brazil refraining from going nuclear, it is more likely to be found within the regional context of the Latin American Nuclear-Weapon-Free Zone, where they would have a higher consciousness of and receptivity to the feelings, desires and

A NUCLEAR-WEAPON-FREE ZONE IN AFRICA

influence of their neighbours, all of whom are developing countries with similar problems, than to the pressures of the rich industrial powers. At any rate, the Treaty of Tlatelolco holds out more hope in this regard than does the NPT, which both countries regard as a discriminatory treaty that the nuclear powers are trying to impose on them.

OUTLINE OF EFFORTS TO CREATE A NUCLEAR-WEAPON-FREE ZONE IN AFRICA

The first proposal for a NWFZ in Africa was made in 1960 after the first nuclear test explosion in the Sahara Desert by France. At that time, eight African countries raised the matter in the United Nations but did not press it. The following year, 14 African states formally proposed in the UN General Assembly a resolution for preventing the extension of the nuclear arms race to Africa and for making Africa a 'denuclearised zone'. The resolution was approved by the General Assembly. It called on all Member States to refrain from conducting nuclear tests in Africa, and from using the area for testing, storing or transporting nuclear weapons, and asked them to consider and respect the continent of Africa as a denuclearised zone. The Soviet Union supported the proposal but the United States and its allies found it unacceptable on the ground that the prohibition of testing meant an uninspected and uncontrolled moratorium.

In July, 1964, at the first Summit Conference of the Organisation of African Unity (OAU), the Heads of State and Government of the African countries issued a solemn declaration on the denuclearisation of Africa and announced their readiness to undertake by treaty not to manufacture or acquire control of nuclear weapons. This declaration was endorsed at a Summit Conference of Non-Aligned Countries held in October of the same year.

In 1965, 28 African states submitted a proposal in the General Assembly to endorse the declaration on the denuclearisation of Africa issued at the Summit Conference of the OAU the previous year. The resolution was overwhelmingly approved by the General Assembly in an almost unanimous vote, including all the nuclear powers except France. In addition to endorsing the declaration on the denuclearisation of Africa, the General Assembly

1. reaffirmed the call on all states to respect the continent of Africa as a NWFZ and to abide by the declaration;
2. called on all states not to use or threaten to use nuclear weapons in Africa;
3. called on all states not to test, manufacture, use or deploy nuclear weapons in Africa or to acquire such weapons or take 'any action which would compel African

A NUCLEAR-WEAPON-FREE ZONE IN AFRICA

states to take similar action',
4. urged the nuclear powers not to transfer nuclear weapons, information or technological assistance to the national control of any state in any form which could assist in the manufacture or use of nuclear weapons in Africa: and
5. expressed the hope that the African states would initiate steps through the OAU with a view to implementing the denuclearisation of Africa.

Nothing further was done to implement the declaration or the resolution. Nine years later, however, in December, 1974, twenty-six African countries again proposed a resolution in the General Assembly which was unanimously adopted by all countries including France, which had some years earlier shifted its nuclear testing from Algeria to the Pacific Ocean island of Mururoa. The resolution reaffirmed the previous resolutions and again called on all states to refrain from testing, manufacturing, deploying, transporting, storing, using or threatening to use nuclear weapons on the African continent.

The subject of a NWFZ in Africa was again discussed by the 30th session of the UN General Assembly in 1975. On this occasion 34 African states, led by Nigeria, introduced a resolution similar to that adopted in 1974, under the agenda item 'Implementation of the Declaration on the Denuclearisation of Africa' supporting the establishment of a NWFZ in Africa. The resolution was again adopted by a unanimous vote, including all the nuclear powers, on December 11, 1975.

At the 1976 session of the General Assembly some new elements were brought into the discussion. The African states were becoming increasingly concerned about the growing capabilities of South Africa in the field of nuclear energy, which gave it the potential of making its own nuclear weapons. The resolution which was adopted by consensus, without a formal vote, dealt specifically with this aspect for the first time. Among its provisions were the following:

> Concerned that further development of South Africa's military and nuclear-weapon potential would frustrate efforts to establish nuclear-weapon-free zones in Africa and elsewhere as an effective means for preventing the proliferation, both horizontal and vertical, of nuclear weapons and for contributing to the elimination of the danger of a nuclear holocaust ...

> Appeals to all States not to deliver to South Africa or place at its disposal any equipment or fissionable material or technology that will enable the racist regime of South Africa to acquire nuclear-weapon capability.

A NUCLEAR-WEAPON-FREE ZONE IN AFRICA

BASIC CONSIDERATIONS CONCERNING NUCLEAR-WEAPON-FREE ZONES

In spite of the growing interest in NWFZs in several areas of the world, none of the various proposals, with the important exception of the South Pacific zone, has yet progressed very far. Nevertheless, as previously indicated, there are fashions in political affairs, and political, military and strategic developments are moving at such a pace that opportunities may exist tomorrow that appear to be impossible today.

The failure of the Rapacki Plan for a NWFZ in Central Europe, and of other initiatives that were no less persistently pursued, was due to the fact that they were aimed at altering existing security arrangements or the existing military balance in the area. On the other hand, the Latin American initiative succeeded because it was a genuine cooperative effort undertaken by the countries of the region in order to keep nuclear weapons out of their area. The initiative was not aimed at any particular country or security arrangement but was perceived by the parties concerned as being in their common interest.

Moreover, the Latin American countries went ahead with their project without requiring the participation of every country in the area. They were content to go forward with establishing the zone in the maximum area possible. If Cuba was not ready to join or the Soviet Union to give support, the Latin American countries were prepared to begin on a smaller scale and to work toward achieving their full goal in due course.

Since the African NWFZ was also conceived as a genuine cooperative effort in the common interest of the African countries, it too has a chance of success. Here too, however, if the zone is to be created in the foreseeable future, it may be necessary for the countries of Africa to proceed without the participation of South Africa and, perhaps, Egypt. Even though the absence of these two countries, which are the most advanced in nuclear technology, would be a short-coming that would reduce the effectiveness of the NWFZ, there would always be the possibility of their joining later.

With respect to some of the other proposals for NWFZs now being actively considered (the Middle East, South Asia, and the Balkans) the situation is entirely different. Since these proposals seem to be more strategically and politically inspired rather than cooperative efforts conceived and worked out in joint consultations by the main countries of the respective areas, the prognosis for them must be considered poor. The situation might, of course, change if genuine peace and minimum conditions of confidence are established between Israel and the Arab states or between India and Pakistan. Until such conditions are created, it is obvious that no country that regards its security as threatened can be politically manoeuvred or pressured into accepting a nuclear-free

A NUCLEAR-WEAPON-FREE ZONE IN AFRICA

status. No country would freely agree to become a party to any treaty, let alone one that could vitally affect its security, unless it considered that doing so was clearly in its interest, or at least not prejudicial to its interests. In fact, it is axiomatic that even if a country becomes party to a treaty, it will not remain a party if it considers that events have changed the basic situation or have caused the treaty to be against its interests.

Nevertheless, the idea of NWFZs is undoubtedly a good one. It provides a means whereby non-nuclear countries can, by their own initiative and effort, ensure their greater security. It can be an effective instrument not only to prevent non-nuclear countries in a given region from 'going nuclear' but also prevent the stationing or deployment of nuclear weapons in the countries of that zone by the nuclear powers. In addition it can be a means of obtaining pledges from the nuclear powers not to use or threaten to use nuclear weapons against any of the countries in the area of the zone. The United States, Great Britain, and the Soviet Union, which refused to include such a pledge in the NPT, have agreed to a commitment of this kind by signing and ratifying Protocol II of the Treaty of Tlatelolco. China and France, which are not parties to the NPT, have likewise become parties to Protocol II.

Moreover, such zones could avoid the discriminatory features of the NPT and provide the mutual security desired by countries that for one reason or another do not want to become parties to the NPT. Thus NWFZs can be a very effective way of promoting and strengthening the non-proliferation regime. The creation of such zones would in no way conflict with the NPT but would provide a means for extending and reinforcing the objectives of the Treaty.

NWFZs could also provide a sound reason and logical basis for promoting the peaceful uses of nuclear energy and for facilitating the receipt of nuclear assistance from the countries that are suppliers of nuclear materials, equipment and technology. If there are valid reasons and any inclination to establish international or regional nuclear fuel cycle centres for enriching uranium, storing or reprocessing spent fuel, or handling nuclear wastes, it would seem that a NWFZ might provide a suitable site and rationale for such centres. The economic and security benefits, and the increased effectiveness of safeguards against diversion or theft of nuclear material attributable to a large-scale regional centre as compared to a number of smaller national facilities, could provide some additional arguments in favour of the creation of such zones. Because the present state of nuclear technology makes a peaceful nuclear explosive device equivalent to a nuclear weapon, it would have to be made clear, in order that nuclear assistance from nuclear suppliers be facilitated, that no

A NUCLEAR-WEAPON-FREE ZONE IN AFRICA

country in the zone could itself conduct any peaceful nuclear explosion.

The problem would remain, and indeed, would become more serious, that as countries acquired nuclear facilities, material, and know-how from their peaceful power programs, they would also acquire the know-how for making nuclear weapons.

There is no solution to the dilemma thus posed except in the creation of a world in which countries have no need and see no advantage in acquiring nuclear weapons, and in which the nuclear-supplier countries adopted practices that reduced, to the extent feasible, the possibilities and temptations for countries to 'go nuclear'. Up to the present time the nuclear superpowers and the members of the 'London Suppliers' Club' have not created such conditions and do not seem likely to do so in the foreseeable future.

Thus the dilemma and the risks of proliferation will continue in the world. Because of the attractions of nuclear power programs, it is probably too late to turn back or even stop the trend towards the spread of nuclear technology, and the risks of proliferation will probably increase. This, in itself, is an argument in favour of supporting the creation of more NWFZs, since such zones would at least reduce the risks that its members would become nuclear proliferators.

It is, no doubt, true that the continuing development and promotion of the peaceful uses of nuclear energy will give a growing number of countries the option of 'going nuclear' if they so choose. But the creation of NWFZs would increase the likelihood of their members remaining 'latent proliferators', without exercising the option to 'go nuclear'. With the perilous prospect of a world of many nuclear powers, any avenues that hold out the possibility of reducing or restraining the trend towards proliferation are worth traversing.

Resolutions of the UN General Assembly are not legally binding and hence it is necessary that the undertakings and provisions for the creation of a NWFZ should be embodied in a legally binding form by treaty.

Provided that the idea of a NWFZ is conceived and promoted by the non-nuclear countries of a given area as a genuine effort to increase the security of all the countries affected and not just some of them, there is no reason additional NWFZs should not be established in different areas of the world. Their establishment would be facilitated if the nuclear powers would demonstrate their wholehearted support for the creation of such zones not only as a means of preventing the further spread of nuclear weapons, but also as a way of furthering the peaceful uses of nuclear energy and of providing security assurances to the countries in the denuclearised zone.

The NPT itself, although it does not explicitly encourage or facilitate the creation of NWFZs, does nothing to discour-

A NUCLEAR-WEAPON-FREE ZONE IN AFRICA

age or detract from them. In fact, it gives them its blessing. The initiative, however, remains with the parties in any region. The NWFZ is one of the easiest and best ways by which the non-nuclear countries can do something by and for themselves. In a number of important respects it provides a more effective means than the NPT for maintaining and strengthening international peace and security.

THE POLITICS OF A NUCLEAR-WEAPON-FREE ZONE IN AFRICA

Very few countries in Africa have the potential of 'going nuclear' in the near future. In fact, South Africa is the only near-nuclear power in Africa, although Egypt must be regarded as a potential nuclear power. South Africa is not a party to the NPT.

South Africa, in addition to having several nuclear reactors (under IAEA safeguards), is one of the largest uranium-producing countries. Moreover, South Africa has announced that it has a new secret process for enriching uranium, in which case it can explode a uranium device without waiting to acquire a plutonium device. The vice-president of the South African Atomic Energy Board stated after the Indian explosion that South Africa had the capability of making a bomb and was more advanced in nuclear technology than India. He stressed that South Africa would use its available uranium and nuclear technology only for peaceful purposes (whatever that may mean after the Indian tests), and that if South Africa should ever decide to 'go nuclear' it would be for what it regarded as overriding military and security reasons.

South Africa has explicitly refused to renounce nuclear defence. In the spring of 1976 Prime Minister Johannes Vorster again stated that South Africa had the technological capability to produce atomic bombs. South Africa is reported to have entered into agreements providing for nuclear cooperation with both France and West Germany. In addition, in May, 1976 South Africa and France announced an agreement for the sale to South Africa under international safeguards of two large power reactors, each with a capacity of about 1000 megawatts. (These reactors could produce enough plutonium to make more than a hundred atomic bombs each year of the same size as the Nagasaki bomb.) Earlier negotiations to purchase the two reactors from a US-Dutch-Swiss consortium had fallen through because of US insistence that it must approve how and where the spent fuel was reprocessed to extract the plutonium.

In June 1976 the Summit conference of the OAU condemned the sale of the reactors as a 'serious threat against the peace and security of Africa and the world'.

A NUCLEAR-WEAPON-FREE ZONE IN AFRICA

The black African nations and their friends also wanted to have South Africa expelled from the IAEA but settled for a request to the Board of Governors to review the annual designation of South Africa as the member of the Board for Africa. They failed in their efforts to have the UN Security Council adopt sanctions against South Africa by cutting off all assistance. They did succeed in having the UN General Assembly adopt a series of resolutions condemning South Africa and the countries that provided it with military assistance. And as mentioned above in the resolution on the African NWFZ as well as in one adopted on November 10, 1976, the UN General Assembly called for the cessation of nuclear cooperation with South Africa. While the emotions and motivations of the black African countries are understandable, one cannot help but wonder whether actions which tend to isolate South Africa and drive it into a corner may not have the effect of strengthening the hand of the hawks in South Africa who want that nation to 'go nuclear'. If South Africa has in fact developed a process for enriching uranium, then it can 'go nuclear' even without importing nuclear supplies and technology.

Egypt signed the NPT in 1968 and ratified it in 1981 despite announcing that it would not become a party unless Israel did. Egypt is far behind Israel in nuclear technology, and stories have circulated in the diplomatic world for some years that Egypt had (without success) asked India and other countries to help it acquire nuclear weapons or nuclear-weapons capability. The agreement in June, 1974 by President Nixon to provide two 600-megawatt nuclear power reactors, one to Egypt and one to Israel, raised serious questions. Each reactor could produce enough plutonium to make more than 10 medium-sized nuclear bombs a year. No matter what safeguards are written into such an agreement, including placing the reactor under IAEA safeguards and returning the spent fuel to the United States for reprocessing, it is always possible for a country wanting to do so to evade its commitments or abrogate the agreement. In answer to those who oppose the supply or sale of nuclear reactors to Egypt and Israel, US officials say that if their country does not go ahead, then France or some other country will do so and will probably not insist on safeguards as strict as those the United States would require.

Due to Israel's opposition to placing all its fissionable material under international or US safeguards and inspection, and the declining fortunes of the nuclear power industry generally, the likelihood of any agreement to supply reactors to Egypt and Israel is unclear. In any case, it is clear that, until relations between Israel and its Arab neighbours improve, both sides will want to keep their nuclear options open.

A NUCLEAR-WEAPON-FREE ZONE IN AFRICA

One of the main problems that has hindered the implementation of the declarations and resolutions on an African NWFZ is that the two African countries most technologically advanced in the field of nuclear energy have been in an anomalous position. Although South Africa can readily 'go nuclear' most of the other African states have instituted a boycott and refuse to have any traffic with that nation because of its policy of apartheid and its reluctance to grant independence to Namibia (South West Africa). At the other end of the continent, Egypt, which is a potential nuclear power (although it is well behind South Africa) has been reluctant to tie its hands in nuclear enterprises unless Israel also does so. The other African states are much less advanced in nuclear technology; none of them can be considered as being a near-nuclear power and very few even as being potential nuclear powers.

Nevertheless, as more African countries acquire nuclear research and power reactors, and with them expertise in nuclear technology, more of them will also become potential nuclear-weapon powers. Some oil-rich states such as Algeria, Libya and Nigeria have demonstrated an interest in obtaining nuclear technology. So, too, has Zaire which has uranium resources. If any of these countries is determined to acquire expertise in nuclear technology and a nuclear-weapon capability, it can do so in about a decade or so. And other resource-rich African states will not be far behind.

It is not in the interest if any African state for any other African state to 'go nuclear'. Any country which acquires nuclear weapons automatically becomes a nuclear target. Moreover, if any African country 'went nuclear' it would reduce the security of all the African states and possibly stimulate a nuclear arms race among them. One can easily envisage such an unfortunate development before the end of this century if any African state decides to acquire a nuclear-weapon capability either by manufacturing a nuclear weapon or a 'peaceful' nuclear explosive device.

Hence, in Africa as in Latin America, where there are many similarities in circumstances favouring the creation of a NWFZ, the time is opportune to proceed with concrete efforts to prepare for the establishment of such a zone. As previously noted it would be quite feasible to establish a NWFZ in Africa without the participation of South Africa.

As was demonstrated at the OAU conference in 1964, and by the unanimous support of the black African and Arab states of the six UN General Assembly resolutions, there is a community of interest in Africa for the creation of a NWFZ. It would not require a great deal of effort to achieve agreement even if the zone did not at first include all the countries of the continent. Moreover, it is likely that the African countries that are not parties to the NPT would become parties to a treaty creating a NWFZ in Africa, particularly if the zone is

A NUCLEAR-WEAPON-FREE ZONE IN AFRICA

established before any black African country becomes a 'near-nuclear' or 'threshold nuclear' power. It would seem to be wise to encourage these regional efforts. Such efforts would have the maximum chance for success precisely because there is no prospect of any of these countries 'going nuclear' in the immediate future or even acquiring a nuclear option. Once a country acquires that option, it becomes much harder to persuade it to give it up.

Moreover, there can be specific security advantages for black Africa, as well as for the continent in general, if the OAU were to undertake determined efforts to establish a NWFZ in Africa. On the other hand, the creation of such a zone might eliminate or at least reduce the incentives for South Africa to 'go nuclear', as this would remove the possibility of its perception of any potential nuclear threat against it from an African country. On the other hand the very fact of a NWFZ in black Africa would help to create a non-nuclear climate and increase international pressures on South Africa to refrain from or resist temptations to 'go nuclear', and would strengthen domestic forces there holding similar views.

The combination of foreign and domestic pressures played a decisive role in changing South Africa's attitude and policies regarding the related problems of Rhodesia (Zimbabwe) and Namibia. Paradoxically, the active dissidence of the blacks in South Africa that began with the Soweto riots on June 16, 1976, can have the effect of retarding rather than speeding up any tendency for South Africa to manufacture nuclear weapons. If South Africa should feel beleaguered and isolated, that might increase internal pressures for South Africa to 'go nuclear'. If South Africa should ever decide to 'go nuclear', it would be because of pressure within the country from the military establishment and political 'hawks' to do so for military and security reasons and in an effort to exploit its possession of nuclear weapons for political blackmail or to deter an attack by black African states or guerrilla forces. On the other hand, if the rioting by the blacks has the effect of increasing international and domestic pressures to intensify the declared policy of detente in South Africa and lead to a speeding up of South African moves toward accommodation with black Africa and its own blacks, the 'doves' in South Africa would be strengthened in their efforts to prevent, or at the least, to delay any decision to 'go nuclear'.

If tendencies toward detente are carefully nurtured and fostered at home and from abroad, the South African detentists have a reasonably good chance of winning the struggle with the hard-liners. Since the value of nuclear weapons in the context of Africa against either organised military forces or guerrillas is very dubious from the military point of view, and execrable in political and moral terms, it is conceivable that South Africa may opt for a nuclear policy similar to that attributed to Israel, namely, to acquire the capability of

A NUCLEAR-WEAPON-FREE ZONE IN AFRICA

producing nuclear weapons very quickly but without actually 'going nuclear' by exploding a nuclear device. It would help to increase the security of the entire continent and promote the cause of nuclear non-proliferation if South Africa should adopt such a middle position or 'grey' area between a non-nuclear and nuclear status and remain a 'latent' nuclear power.

It may not be easy for the African states, because of their other preoccupations in southern Africa, to appreciate that their taking the initiative to create a NWFZ could help their military as well as political position as against South Africa. However, their perceptions of their long-term interests outweighed their short term inclinations when they were dissuaded from demanding the exclusion of South Africa as an observer at the NPT Review Conference in May, 1975 and from 'going all out' to expel it from the IAEA in 1976.

The establishment by the black African states of a NWFZ might very well generate new moral pressures in South Africa and throughout the world against South Africa's acquiring nuclear weapons. One need only recall the moral pressures created by India on Great Britain and in the world as a result of India's policies, as exemplified by Gandhi and Nehru, of developing a posture of moral superiority. Whether the creation of a NWFZ in Africa might in some measure have a similar effect is debatable, but nothing would be lost and something might be gained by it.

It is of course possible that, irrespective of what the black African states do about a NWFZ in Africa, internal developments in South Africa or the dynamics of confrontation between black Africa and South Africa, will cause South Africa to 'go nuclear' either by exploding a weapon, or, more likely, a 'peaceful nuclear device' to obtain more leverage in its global and regional international relations. But the creation of a NWFZ in Africa could not under any conceivable scenario increase or hasten any desire or tendency for South Africa to 'go nuclear', and it might have the effect of reducing or slowing down any such tendency. Any time gained in this regard could be used to promote possible peaceful progress between the blacks and whites and, perhaps, even postpone indefinitely the decision of South Africa to become an active proliferator.

Moreover, as indicated earlier and entirely apart from the question of South Africa and Egypt, it would clearly be in the interest of all the black African states to eliminate or reduce the prospects that any of the other states in Africa might one day 'go nuclear'.

In any case, as is now customary, in the case of all arms control agreements, including the Treaty of Tlatelolco and the NPT, any treaty creating a NWFZ in Africa would in all probability contain a withdrawal clause. Hence, if South Africa or any other African state should one day 'go nuclear',

A NUCLEAR-WEAPON-FREE ZONE IN AFRICA

any other state that was a member of the zone could withdraw from the zone if it felt that its security was jeopardised.

In addition, an African NWFZ could provide the framework and the institutional machinery for facilitating and promoting the peaceful uses of nuclear energy in Africa in much greater measure than do the IAEA and other existing agencies for promoting the technological development of the developing countries.

Above all, by ensuring the total absence of nuclear weapons from their territories, the states of Africa that became parties to the zone would also facilitate, to the extent possible, their removal from the rivalries and nuclear threat of the nuclear powers. The zone could be the instrument for their obtaining pledges of non-use of nuclear weapons against them and the means of resisting attempts by the nuclear powers to acquire nuclear bases or other similar special privileges in the area which could derogate from the independence or freedom of action of the countries concerned.

Thus, there are few disadvantages and many advantages that would accrue to the nations of Africa and, indeed, to all the world from the establishment of a NWFZ in Africa.

PROPOSALS

On several occasions the black African states have initiated unanimously adopted resolutions in the UN General Assembly calling for the recognition of the continent of Africa as a NWFZ. It is time to transform the General Assembly resolutions into a binding treaty.

In this regard, it might be well if the African states were to follow the precedent of the Latin American experience. The Treaty of Tlatelolco was initiated in April, 1963, after the Cuban missile crisis, when five Latin American Presidents called for the denuclearisation of the area, a call which was supported by the UN General Assembly in the fall of that year. The following year the states of Latin America, at the invitation of Mexico, met in conference and established a Preparatory Committee for the Denuclearisation of Latin America. It was this Preparatory Committee, under the dedicated Chairmanship of Alfonso Garcia Robles, (then Under-Secretary for Foreign Affairs and later Foreign Minister Minister of Mexico) and with some assistance from the author (who was appointed by UN Secretary-General U Thant to be Technical Consultant to the Commission) that prepared the draft of what became the Treaty of Tlatelolco. The African states have already obtained the blessing and support of summit conferences of the OAU and of the Non-aligned Countries for the denuclearisation of Africa. What is required now is that some dedicated black African leader become the 'father' of an African treaty, as Garcia Robles was of the

A NUCLEAR-WEAPON-FREE ZONE IN AFRICA

Latin American Treaty, and obtain the support of his government to pressing that objective.

The first step is to have the OAU, or a group of African states, or even a single government, sponsor a conference to establish a Preparatory Committee for a Nuclear-Weapon-Free Zone in Africa consisting of all the African states that are willing to participate. It would probably be better if the 40 or so African states followed the Latin American example and organised their own conference and Preparatory Committee, which would remain under their own control free of the politics, pressures and rivalries of the entire membership of the United Nations. The nuclear powers and other interested states might be invited as observers rather than as participants, to follow the work of the Committee. In this manner the continuing interest of such nations could be maintained and they could contribute to the progress of the work. But it would not be necessary to obtain their approval for each step.

Once such a Committee is created under the active leadership of a widely respected African personality, half the struggle would be won. The African Committee could call on the Secretary-General of the United Nations for assistance, as did the Latin American Committee. It could also call on OPANAL, the organisation set up by the Treaty of Tlatelolco to supervise its implementation and operation. And it could no doubt take advantage of the support, the facilities and the Secretariat of the OAU. Annual or periodic reports could be made to the United Nations and the work of the Committee could be discussed and supported by the General Assembly.

The task of the Committee would be to elaborate the text of a treaty for the creation of a NWFZ in Africa and to reconcile and harmonise the likely divergent views. This task will not, of course, be simple or quick. But with resolve and patient and skillful leadership this goal can be achieved. As indicated previously, it would not be necessary to obtain the adherence of all African nations from the beginning, although some minimum would be required.

The fundamental objective has already been set by the African states and the road to the goal has already been pioneered by the Latin Americans. All that is now required is the necessary political leadership.

A NUCLEAR-WEAPON-FREE ZONE IN AFRICA

APPENDIX

UN General Assembly Resolution A/Res. 31/69 adopted on 10 December 1976

THE GENERAL ASSEMBLY

Implementation of the Declaration on the Denuclearisation of Africa

Recalling its resolutions 1652 (XVI) of 24 November 1961, 2033 (XX) of 3 December 1965, 3261 E (XXIX) of 9 December 1974 and 3471 (XXX) of 11 December 1975, in which it called upon all States to consider and respect the continent of Africa, including the continental African States, Madagascar and other islands surrounding Africa, as a nuclear-weapon-free zone,

Recognising that implementation of the Declaration on the Denuclearisation of Africa would contribute to the security of all the African States and to the goals of general and complete disarmament,

Bearing in mind that the Assembly of Heads of State and Government of the Organisation of African Unity at its thirteenth ordinary session, held at Port Louis from 2 to 6 July 1976, expressed grave concern over the continuing collaboration between certain States Members of the United Nations and the racist regime of South Africa, particularly in the military and nuclear fields, thereby enabling it to acquire nuclear-weapon capability,

Concerned that further development of South Africa's military and nuclear-weapon potential would frustrate efforts to establish nuclear-weapon-free zones in Africa and elsewhere as an effective means for preventing the proliferation, both horizontal and vertical, of nuclear weapons and for contributing to the elimination of the danger of a nuclear holocaust,

1. Reaffirms its call upon all States to respect and abide by the Declaration on the Denuclearisation of Africa;

2. Further reaffirms its call upon all States to consider and respect the continent of Africa, including the continental

A NUCLEAR-WEAPON-FREE ZONE IN AFRICA

African States, Madagascar and other islands surrounding Africa, as a nuclear-weapon-free zone;

3. Appeals to all States not to deliver to South Africa or place at its disposal any equipment or fissionable material or technology that will enable the racist regime of South Africa to acquire nuclear-weapon capability;

4. Requests the Secretary-General to render all necessary assistance to the Organisation of African Unity towards the realisation of its solemn Declaration on the Denuclearisation of Africa, in which the African Heads of State and Government announced their readiness to undertake, in an international treaty to be concluded under the auspices of the United Nations, not to manufacture or acquire control of nuclear weapons;

5. Decides to include in the provisional agenda of its thirty-second session the item entitled 'Implementation of the Declaration on the Denuclearisation of Africa'.

Chapter Nine

NUCLEAR-FREE ZONES: PROBLEMS AND PROSPECTS

Ken Coates

The United Nations Special Session on Disarmament, convened in New York in 1978, excited world opinion by reaching agreement on the goal of 'general and complete disarmament'. This goal was deliberately coupled with a series of intermediate objectives. One of these concerned the establishment of nuclear-weapon-free zones. Article 33 of the final statement of the 1978 Special Session states that the establishment of such zones 'on the basis of agreements or arrangements freely arrived at among the states of the zone concerned, and the full compliance with those agreements or arrangements, thus ensuring that the zones are genuinely free from nuclear weapons, and respect for such zones by nuclear-weapons states, constitutes an important disarmament measure'.

Later the declaration goes on to spell out this commitment in a little more detail. It begins with a repetition:

> The establishment of nuclear-weapons-free zones, on the basis of arrangements freely arrived at among the states of the region concerned, constitutes an important disarmament measure,

and then continues:

> The process of establishing such zones in different parts of the world should be encouraged with the ultimate objective of achieving a world entirely free of nuclear weapons. In the process of establishing such zones, the characteristics of each region should be taken into account. The states participating in such zones should undertake to comply fully with all the objectives, purposes and principles of the agreements or arrangements establishing the zones, thus ensuring that they are genuinely free from nuclear weapons.

> With respect to such zones, the nuclear-weapon states in turn are called upon to give undertakings, the modalities

of which are to be negotiated with the competent authority of each zone, in particular:

a. to respect strictly the status of the nuclear-free zone;
b. to refrain from the use or threat of use of nuclear weapons against the states of the zone...

States of the region should solemnly declare that they will refrain on a reciprocal basis from producing, acquiring, or in any other way, possessing nuclear explosive devices, and from permitting the stationing of nuclear weapons on their territory by any third party and agree to place all their nuclear activities under International Atomic Energy Agency safeguards.

Of course, the final document of the Special Session is not beyond criticism. It is quite possible to identify a number of problems to which the document gives no attention, or inadequate attention. These become apparent even in the discussion about the establishment of nuclear-free zones. But that discussion itself owes a very great deal to the fact that agreement was reached in New York in 1978 and although no-one would suggest that there now exists a panacea for controlling and reversing the arms race, nonetheless the Special Session has offered us a groundwork for the development of nuclear-free zones.

There are two categories of involvement in nuclear-free zones. First, the states constituting any such zone need to reach an agreement about the commitments which are involved in nuclear-weapon-free status, and to resolve upon means for jointly enforcing that status. Second, the existing nuclear powers outside such a zone need to be brought to an accord which can underwrite it, by promising respect for its integrity. If the nuclear powers refuse agreement to respect the zone, then it will not succeed in establishing itself as much more than a propaganda commitment. Important though such a commitment undoubtedly is, both the First UN Special Session and the modern peace movement have expected greatly more than this. It is, of course, always possible that the nuclear powers could ratify an agreement concerning such a zone, and then violate their own obligations. This possibility raises very important problems, to the discussion of which we shall return.

THE AGREEMENT BETWEEN MEMBERS OF A NUCLEAR-FREE ZONE

Obviously every state which seeks the establishment of a nuclear-free zone is seeking to share in an agreement prohibiting the ownership, construction, acquisition, or use of

NUCLEAR-FREE ZONES: PROBLEMS AND PROSPECTS

nuclear weapons. Such an agreement may find problems of definition in determining what constitutes a nuclear weapon. Do 'nuclear weapons' mean weapons employing nuclear explosives, or can they include weapons deriving their propulsion from nuclear fuels? Let us assume that this argument can be resolved: then we immediately require a whole series of definitions to control the meaning of other terms of the agreement. What is 'ownership'? What is 'construction'? 'Acquisition'? Even, 'use'?

Obviously the possession of nuclear weapons can involve a variety of real situations. Some states own their own weapons, while others have developed a number of different relationships promoting the deployment of weapons belonging to others. The treaty for a nuclear-free zone will need to cast its net wide enough to forbid all of them equally. Weapons deployed directly or indirectly on behalf of a third party are, for the purposes of such an agreement, perhaps even worse than weapons directly owned and completely controlled by the party concerned. Each link in the deployment chain is liable to give rise to an uncertainty about use. A nuclear-free zone will obviously prohibit the placement of nuclear weapons in its agreed area, but most participants would presumably seek also to prohibit their co-signatories from owning and deploying such weapons in territories outside the agreed boundaries of the zone. Although this reservation seems far-fetched, there are several practical cases in which it is immediately relevant.

For example, some countries straddle boundaries. Turkey, for instance, is one. If Turkey were to seek to become a member of a European nuclear-free zone, would Turkish territory in Asia Minor be included? If it were not, consenting parties to the treaty inside Europe could imaginably be bombarded by Turkish-based nuclear weapons held outside the agreed boundaries of the zone. It seems fairly clear that no state could be allowed to occupy a schizoid position of this kind. The example of Turkey is not the only one in this category, and that is why the Soviet Union cannot simultaneously place part of its territory in, and part outside, a nuclear-free zone. This fact, not any lack of sensitivity to the 'European' status of Russia, persuades many advocates of European Nuclear Disarmament of the advantage of the slogan of a 'Europe free of nuclear-weapons, from Poland to Portugal' rather than 'from the Atlantic to the Urals'.

A nuclear-free zone treaty would also have to define manufacture or construction in a clear and precise way. Every nuclear reactor is a potential hazard in this respect. If British nuclear power stations have supplied plutonium for the American nuclear weapons programme, then surely they are helping the 'manufacture' of nuclear weapons. The British authorities deny such co-operation, saying that the plutonium which they have supplied to the United States was for civilian purposes only. But if the civilian use of British plutonium liberates American produced plutonium for the American military programme, is this not a specious quibble? States consenting to join a nuclear-free zone may wish to elaborate a framework which could clarify policy about such matters.

Let us assume that precise definitions of all these terms can be agreed. No signatory will make nuclear weapons or conduce to their manufacture, either on their own behalf, or in co-operation with any third party; no signatory will allow deployment of nuclear warheads or any other nuclear explosive machines, either on their own behalf or in co-operation with any third party, and no signatory will permit the transportation of nuclear weapons over their territory, including their territorial waters and their national air space.

It would be both reasonable and simple for the parties consenting to such a treaty to further pledge themselves to refrain from testing nuclear devices outside the territorial area covered by the treaty agreement.

Although it would not be by any means useless for existing non-nuclear powers to league themselves into nuclear-free zones, we may assume that some nuclear-free zones in Europe would be likely to include states which are at present deploying nuclear weapons of one kind or another within their national territories. If we could suppose that say, Austria, Switzerland and Yugoslavia were able to establish a nuclear-free zone treaty, this would undoubtedly exercise some persuasive influence in the other parts of Europe which are less fortunate in that their territories are already defiled by the presence of nuclear weapons. But if there is an agreement in the Central European zone or in the Balkans, either case will involve the physical withdrawal of nuclear weapons which are presently emplaced. Rather precise arrangements will be necessary to determine such withdrawal.

At the same time, members of a nuclear-free zone will need to decide how far they wish to extend their prohibitions to cover equipment which is ancillary to the deployment of nuclear weapons. Should launchers and other emplacements also be specifically forbidden within the agreed territory? It would seem so.

THE UNDERWRITING OF NUCLEAR-FREE ZONES BY THE NUCLEAR POWERS

Naturally, nuclear-free zones offer but the sparsest imitation of security if they are not recognised by the existing nuclear powers and if those powers do not give solemn undertakings to respect their status.

In the case of the Treaty of Tlatelolco, there are two protocols which have been endorsed by the nuclear powers,

NUCLEAR-FREE ZONES: PROBLEMS AND PROSPECTS

respecting and accepting the agreement that nuclear weapons may not be deployed within the Treaty area. The second protocol, for instance, agrees to fully respect the Treaty, and to refrain from violating it. It further commits its signatories not to use or threaten to use nuclear weapons against the contracting parties of the main Treaty.

Many people have questioned whether the Latin American Treaty goes far enough in the requirements it imposes on the nuclear powers. Should not those powers agree to prohibit the sale of nuclear materials which could be used for military purposes? Should they not agree to disconnect any linkages with nuclear weapons aspects of any existing military alliances? But undoubtedly the major problem about nuclear power recognition of nuclear-free zones is that of enforcement. To this we shall return later.

THE DEVELOPMENT OF A NUCLEAR-FREE ZONE IN EUROPE

The two sub-regions of Europe which have made most progress towards the development of nuclear-free zones are the Northern (Nordic) area and the Balkan zone. A recent conference in Athens brought together representatives from Romania, Bulgaria, Yugoslavia and Turkey, at the invitation of the Greek Government which hosted the meeting. (Interestingly, there was a parallel conference involving some of the peace movements of Europe.) Although this failed to agree a communique, it afforded useful opportunities for discussion. The Russell Foundation was denied admission, but we welcome the fact that the gathering was convened, and that the Greek authorities played so prominent a part in it.

The conference was preceded by a joint appeal against the deployment of nuclear missiles in Europe, launched by President Ceausescu of Romania and Andreas Papandreou, the Greek Prime Minister. In a courageous letter to Presidents Reagan and Andropov the two European statesmen opposed the deployment of intermediate range nuclear missiles in Europe, to the considerable annoyance of some client states such as Great Britain. (The Greek Ambassador in London was informed by the Foreign Office that it was 'extremely annoyed' that Greece had failed to consult Britain before joining in so unusual a declaration.)

The Athens conference began after a two week delay, whilst the Turkish Government manoeuvred about whether to join or not. The Romanians in particular reasonably insisted that a Balkan meeting needed Turkish representation, although the Turks were, at best, lukewarm about the proposal. From the beginning they made it clear that they did not believe that the Balkans could be separated from the rest of Europe on matters of nuclear weapons. However, they were anxious to be present in any multilateral gathering of Balkan

NUCLEAR-FREE ZONES: PROBLEMS AND PROSPECTS

states, and they agreed to attend provided the de-nuclearisation plan was placed last on an agenda of five points. Surprisingly, very prompt agreement was reached among the other states to accept this proviso. However, the Turks were not mollified by this, and when the talks took place, they maintained the unconstructive view that 'the right forum for discussing nuclear weapons' control is the US-Soviet talks in Geneva, not somewhere on the periphery'. Since the forum in question subsequently collapsed, most people are hopeful that the periphery may prove more sensitive to disarmament needs than the alleged 'centre' has shown itself. Greece and Turkey are the only Balkan nations harbouring nuclear weapons, so clearly it is necessary to try to secure Turkish agreement to the creation of a nuclear-free zone. Paradoxically, both Greece and Turkey are harbouring American nuclear weapons, even though mutual suspicions ensure the most uneasy of relationships between them. It is arguable that the conflict between Greece and Turkey is perhaps the most intransigent of all the national rivalries on the European mainland, and it is instructive that this fracture exists <u>within</u> one of the alliances, not across the divide of the cold war.

The commitment of the Greeks to a nuclear-free zone is evident, and is made plain in the fact that it was their initiative to convene the February 1984 conference. Bulgaria and Romania have also declared in favour of a nuclear-free zone, as has Yugoslavia. However, the Yugoslavs could reasonably insist that such a zone, constituted in a region so sensitive to superpower confrontations, would be meaningless without fully adequate guarantees from the United States and the Soviet Union. The Athens diplomatic conference ended with a cautious statement, promising to continue discussions.

The problem of securing compliance from the superpowers is, of course, to put it very mildly, no less intractable than that of reconciling the conflicting interests of Greece and Turkey. The United States has expressed strong opposition to the Balkan nuclear-free zone on the grounds that it would alter the strategic balances in the Eastern Mediterranean, without securing adequate concessions from the Soviet Union. The claim is that Warsaw Treaty armed forces are superior in conventional weaponry and that the Turkish and Greek nuclear weapons provide a necessary rung in the scale of 'deterrent' forces. Many of us have argued against this point of view elsewhere, and it is hardly necessary to repeat those arguments here.

More seriously, since the deployment of Soviet intermediate range missiles in the German Democratic Republic and Czechoslovakia, it is argued by some independent scholars as well as Western specialists that the 'neutralisation' of South Eastern Europe disproportionately weakens NATO's Southern flank. If the American presence were to be reduced or removed from this zone, such people have argued that Soviet

NUCLEAR-FREE ZONES: PROBLEMS AND PROSPECTS

pressures could easily increase, not only on Greece and Turkey, but also on Yugoslavia. Clearly, the reciprocal Central European deployments have hindered progress in the Balkans.

The heightened nuclear face-off between the two Germanies and in Central Europe has also been an undoubted set-back to the progress of the Nordic nuclear-free zone.

By contrast with the Balkan area, none of the Nordic states harbours nuclear weapons, and all have signed the non-proliferation treaty. Finland and Sweden are committed neutrals, and the Swedes have a consistent policy of 'non-alignment in peace aiming at neutrality in war'. Denmark, Norway and Iceland are aligned in NATO, but both the Danes and the Norwegians have declared that they will not allow nuclear weapons to be stationed on their territories during peacetime. In one sense, then, there is already an embryonic nuclear-free zone in Northern Europe. However, the separate commitments of the different states of the region do not have the benefit of underpinning by reciprocal commitments on the part of the nuclear powers. The quickest progress towards this would probably come from the formal conclusion of a treaty in the area which would leave nuclear powers caught in the crossfire of world public opinion.

Within the language of 'mutual and balanced force reductions', it would be impossible for either superpower to justify heightened escalation in the central zone of Europe whilst both were moving to disengagement in the North and South. But it will not, long term, be sufficient for one superpower to sponsor the proposed lesser nuclear-free zones and underwrite them alone. This fact will be a brake on development unless there arises a real possibility of loosening up within one or another, if not both, alliance systems. For this reason, the campaign against the deployment of intermediate-range missiles across Europe retains a vital importance. From the point of view of the peace movements, to be sure, the triumph of a small nuclear-free zone on any sector of the dividing line between the blocs would be an important breakthrough. It is increasingly difficult to see any likely circumstances in which a single agreement could embrace the whole of Europe at one sweep: all the tangible progress towards discussion between Governments has so far taken place on a much smaller scale. Yet, even if the sometimes large obstacles to practical agreement in either the Balkan or Nordic regions could all be overcome there would remain a vast problem for the realisation of a comprehensive European solution. Smaller nuclear-free zones will meet with the same difficulties as larger ones, and foremost amongst these is the question of enforcement.

NUCLEAR-FREE ZONES: PROBLEMS AND PROSPECTS

ENFORCEMENT

The enforcement of a nuclear-free zone may not be easy for the individual consenting states within it. It is presumably possible for them to reach agreement about their own conduct, and any mechanisms for inspection and control which may be necessary. There exists an international body which is charged with controlling nuclear materials: the International Atomic Energy Agency. This Agency does not command universal support, and has been open to various criticisms. There seems to be no special reason why the IAEA should be the monopoly custodian of inspectors' powers within nuclear-free zones. The Treaty of Tlatelolco has an enforcement committee. It would be possible for new treaties to agree open inspection mechanisms which were as rigorous as their consenting parties felt to be desirable. However, whilst internal inspection by agreement is simple in principle, enforcement by observance from outside nuclear powers is far more difficult. When nuclear powers feel their interests to be in jeopardy, they have already shown themselves to be less than scrupulous about the observance of international treaties. None of the wars which have plagued humanity during this century would have been possible if all the powers involved in them had respected all the treaty undertakings they had ever signed. More: there is even some evidence that the specific treaty establishing the nuclear-free zone in Latin America has already been violated by one, possibly by two existing nuclear powers.

During the conflict over the Falkland/Malvinas Islands, the British Government sent its fleet to the South Atlantic in order to recover control over these territories. This fleet was apparently not divested of its large arsenals of nuclear depth charges and other weapons. Indeed, there is a good deal of evidence that it would have been impossible so to have divested it, since its 'deterrent' capacity, vis a vis the Soviet Union would have been completely negated for the duration of the war with Argentina, had this been done. Be that as it may, the Falkland flag ship was subsequently compelled to limp around the Pacific, denied admission to ports in both Australia and Japan, on the grounds that it had a stock of nuclear weapons, which are forbidden in those areas. If the Australian Government, which is a key member of the British Commonwealth, felt that it had sufficient evidence to prohibit the dry-docking of this ship, this must seem a prima facie reason for suspecting a serious breach of the Treaty of Tlatelolco during the Falklands war.

In a similar sense, the presence of the American fleet off Nicaragua, in a large-scale deployment covering both the Pacific and Atlantic coasts of that country, also raises the fear that the United States may not be honouring its Treaty obligations.

NUCLEAR-FREE ZONES: PROBLEMS AND PROSPECTS

Of course, both major powers frequently send their fleets into places where they are not entirely welcome. There have been repeated encroachments by Soviet submarines into Swedish territorial waters, for instance. But the breach of a nuclear-free zone agreement should surely be seen as a far more serious dereliction than simple trespass. That apparent violations of the Tlatelolco Treaty can pass with so little public comment is, perhaps, the chief weakness of the nuclear-free zone strategy of peace movements and non-aligned states. Unless we can mobilise effective and damaging public criticism of any and all violations, we shall not succeed in sanitising any great area of the world for any material length of time.

THE ROLE OF THE PEACE MOVEMENT IN FURTHERING DISARMAMENT

If these arguments are justified, the enforcement of nuclear-free zones is only partly a matter for their member states. Of course, adequate inspection and technical control is extremely important, and mechanisms for imposing high standards remain to be achieved. There is much diplomatic work to do, in order to develop the Latin American prototype of a nuclear-free zone so that it could operate in more contentious regions. But what happens when inspection reveals that it is a nuclear power which is acting in violation of non-nuclear agreements? The latest evidence on this score is provided, in a related sphere: the United States' Government has been credibly accused of mining Nicaraguan territorial waters, and causing damage both to Soviet and Japanese ships, amongst others. Clamorous evidence was offered by American politicians to the effect that mines were directly placed by the Central Intelligence Agency. The refusal of the United States' Government to answer Nicaraguan complaints to the World Court, and its subsequent disregard for the Court's findings, show us some of the limits of international law, when we confront such crucial issues as nuclear disarmament. How may the weak compel the strong? It seems transparently plain that institutions for reaching public opinion are more important than institutions of force, if this task is not to be rendered totally futile.

Links between peace movements have been the first stage in diagnosing this problem. Already the Japanese Council Against A and H Bombs has pioneered international contact on the widest scale and reached out to form links with the growing campaign for European Nuclear Disarmament.

In Europe, Conventions of the main peace movements in Western and neutral states have established a continuing dialogue between the main organisations in the field. The first Convention, in Brussels, in 1982, was fortunate in having the

NUCLEAR-FREE ZONES: PROBLEMS AND PROSPECTS

presence of the Venerable Gyotsu Sato, who established close contacts with many of the most important organisations. Strong European representation resulted at the World Conference Against A and H Bombs in Tokyo in 1982, and again in 1983. A powerful delegation from Japan also attended the second European Nuclear Disarmament Convention which was held in Berlin in 1983. The growth of these contacts is of extraordinary importance. More and more American peace movements are relating to the initiatives of both European and Pacific peace activists. If there is one further task which now presses all of us, I think it is this: in a number of continental areas, the debate on the threat of nuclear war lags far behind the level already reached in the European and Pacific 'theatres'. I think that together, we should seek to consider how to encourage that debate, and how to widen the circle of contacts, so that the peace movement begins to create a veritable unity of all the peoples, able to defend all real steps towards disarmament, and to oppose any regression, no matter by whom. The forums of peace movements, both in the Pacific and in Europe, understand the need for a policy of strict non-alignment. Paradoxically, peace movements are less organised and less extensive in many of the areas where the movement of non-aligned states is strongest. Generally it is these states which suffer most from the bad behaviour of nuclear powers. Somehow, surely we must find a way to join our forces to this question.

CONCLUSION

Gordon Thompson

My colleague David Pitt closes his introduction to this volume by framing the nuclear arms issue in terms of the perspectives of sheep and of dragons. This is an attractive analogy, given the reality of the world's nuclear arsenals. A few governments, primarily those of the Soviet Union and the United States, control nuclear weapons and delivery systems which could devastate much of humanity in a few hours. Against the power of these 'dragons', many governments and citizens of nations outside the nuclear club, and not a few citizens of the nuclear nations themselves, must inevitably feel like 'sheep'.

Our contributors describe some of the achievements and aspirations of those sheep who long to be free of the nuclear threat. By declaring their territories to be nuclear-weapon-free zones, these sheep hope both to prevent nuclear arms races in their own regions of the world and to curb the global arms race. The essays collected here show that there has been some success in preventing regional nuclearisation, but up to now little direct effect on the worldwide nuclear arms race.

Readers who wish to learn some of the details of that arms race are commended to a book by William Arkin and Richard Fieldhouse (Nuclear Battlefields: Global Links in the Arms Race, Ballinger Publishing Company, Cambridge, Massachusetts, 1985). This book summarises what is publicly known about the nuclear forces of the United States, the Soviet Union, Britain, France and China, and about the global infrastructure which supports those forces. The latter aspect is particularly interesting: Arkin and Fieldhouse show that each nuclear-weapon state maintains supporting facilities in a variety of other nations, some of which are nominally nuclear-weapon-free. Also, these authors show that six NATO countries and four Warsaw Pact countries, each supposedly 'non-nuclear' by virtue of being a party to the Non-Proliferation Treaty, have numerous delivery systems certified to fire nuclear warheads. At a time of crisis or war, these 'crypto-

CONCLUSION

nuclear' nations will be provided with weapons by their nuclear-armed allies.

Against this background, it is not surprising that the advocates of nuclear-weapon-free zones have had to struggle against great odds. The five nuclear-armed nations are involved in a mighty struggle to protect their supposed 'interests' by threatening to incinerate each other, and many other nations have been drawn into the struggle. In the three instances where large parts of the earth's surface have been declared free of nuclear weapons - Antarctica, Latin America and the South Pacific - either the zones were not previously involved in the nuclear confrontation or the treaties have been written so as to permit that confrontation to continue. Thus, neither the Tlatelolco nor Rarotonga treaties interfere with the transit of nuclear weapons through their respective zones.

At the same time, the entire difficulty of creating nuclear-weapon-free zones cannot be laid at the door of the nuclear-armed nations. Peri Pamir shows that deeply rooted divisions between the Balkan states are a major obstacle to the creation of such a zone in that region. Likewise, a denuclearised Middle East is hard to imagine under present conditions, especially since it is widely assumed that Israel has a small nuclear arsenal. Indeed, the East-West confrontation can mask a variety of other, often quite bitter, regional conflicts.

Despite all the difficulties, our contributors show that the nuclear-free passion is alive and strong in the breasts of the sheep. The example of the Rarotonga Treaty, supplemented by New Zealand's more stringent action of banning port visits by nuclear-armed ships, is likely to engender further moves elsewhere. The potential importance of this trend in regard to the global arms race can be gauged from the vigorous objection of the United States government to New Zealand's action. Although New Zealand's Prime Minister Lange has bent over backwards to assure the American government that New Zealand is not trying to incite similar actions by other nations, political activists around the world must be studying New Zealand's example with acute interest.

Decades of political struggle lie ahead before the concept of nuclear-weapon-free zones can enjoy its full fruition. If that struggle is to succeed, the concept must be employed in a manner which enhances the security of all nations, and which is not seen as simply undercutting the security of any particular alliance. Many issues need to be subjected to careful analysis, and no doubt a large body of such analysis will develop.

We hope that this book will be a useful contribution.

EDITORS AND CONTRIBUTORS

ALEXANDRE BERENSTEIN is President of the Geneva International Peace Research Institute.

KEN COATES is a member of the Bertrand Russell Peace Foundation and Joint Secretary of the International Liaison Committee For European Nuclear Disarmament.

WILLIAM EPSTEIN is Fellow at the United Nations Institute for Training and Research (UNITAR) and formerly Director of the Disarmament Division of the UN.

GREG FRY is at the Strategic and Defence Studies Centre, Australian National University.

HYAM GOLD is at the University of Otago, New Zealand.

PERI PAMIR is at the Graduate Institute of International Studies, Geneva.

DAVID PITT is Charge de Recherche at the Geneva International Peace Research Institute and Consultant to the International Union For the Conservation of Nature.

ALFONSO GARCIA ROBLES is a former Ambassador of Mexico.

RAMESH THAKUR is a Senior Lecturer at the University of Otago, New Zealand.

GORDON THOMPSON is Executive Director of the Institute for Resource and Security Studies, Cambridge, Massachusetts.

SPONSORING ORGANISATIONS

Institute for Resource and Security Studies
27 Ellsworth Avenue, Cambridge, Massachusetts 02139, USA

The institute is a centre for research and public education on the efficient use of resources, protection of the environment, and the furtherance of international peace and security. At present, its work is focused on issues relating to the supply and use of energy and to the proliferation of nuclear weapons.

Financial support for the institute's work comes primarily from grants and contracts from government agencies and public interest organisations, and from gifts by foundations and individuals.

The institute's programs are chosen for their relevance to current policy issues. Research results are disseminated through a public education effort which has involved lecturing, publication through a variety of channels, conference sponsorship, and interviews from print and electronic media.

A program of particular relevance to the present book is the Proliferation Reform Project. This program addresses the problem of nuclear proliferation by exploring and promoting solutions which involve a uniform code of behaviour for all nations. An international group of distinguished experts serves as an Advisory Board to this project.

Incorporated in 1984, the institute is non-profit and tax-exempt. Its affairs are governed by a five-member board of Massachusetts residents, each of whom is active in areas related to the institute's interests.

Geneva International Peace Research Institute
41 rue de Zurich, CH 1202 Geneva, Switzerland

The institute was established in 1979-1980 as an Association under Article 60 of the Swiss Civil Code and in 1984 was reformed as a Foundation (article 80). It is also linked to the pro GIPRI Association (The association supporting GIPRI)

SPONSORING ORGANISATIONS

whose president is Leonard Massarenti. The International Research Institute undertakes studies (both natural and social sciences) in areas relating to security and the promotion of peace; GIPRI is a totally independent organisation sponsored by a Comité d'honneur initially comprising: Denis de Rougemont, Prince Sadruddin Aga Khan, Frank Barnaby, Elise Boulding, Robert Jungk, Manfred Lachs, Sean MacBride, Alfred Kastler, Alva Myrdal, Aurelio Peccei, Max Petitpierre, Hans Ruh, Nagendra Singh, Yoshikazu Sakamoto and Viktor F. Weisskopf. GIPRI is directed by the Foundation Council composed of: Alexandre Berenstein (President), Jean-Pierre Stroot (Vice-president), Ivo Rens (Secretary), Samuel Claude (Treasurer) and André Chavanne, Jacques Diezi, Edouard Kellenberger, Charles-A. Morand and Guy-Olivier Segond.

GIPRI maintains contacts with SIPRI (Stockholm International Peace Research Institute) and other similar Peace Institutes. GIPRI is a member of the International Peace Research Association. GIPRI is supported by the City of Geneva and has close links with the University of Geneva.

INDEX

Africa 5, 114ff
Albania 94
  see also Balkans
Albrecht, U. 87
Algeria 121
Angola 84
Antarctica 7
  signatories to treaty 7
  articles 8
Antunes, E.M. 87
ANZUS 47, 52
  see also South Pacific
Argentina 2, 38
  see also Tlatelolco
Arkin, W. 138
Arlott, J. 86
ASEAN Nations 58
Australia 3, 23ff, 40, 46-66
  see also South Pacific
Austria 67

Bahro, R. 87
Balkans 5, 82, 94ff
Baruch Plan 110
Belgium 85
Bessarabia 96
Bourdet, C. 86
Bragg, M. 86
Brandt, W. 70
Brazil 2, 38
Brockway, Lord 80, 86
Bulgaria 94

Cambodia 78
China (Peoples Republic) 2, 6, 33

Committee on Disarmament (UN) 2
  see also United Nations
Comprehensive Nuclear-Free Zones 60
Cook Islands 26
  see also South Pacific
COPREDAL 10, 16
Cortes, J.R-G. 86
Cuba 116, 123
Curto, F.M. 87
Czechoslovakia 2, 81

Davies, P.M. 86
Davis, T. 34
Duff, P. 86

Egypt 116, 119
Eklund, S. 13
Europe 67-94
European Nuclear Disarmament Movement 85
  see also Balkans

Falklands 135
Favilli, G. 86
Fieldhouse, R. 138
Fiji 25
  see also South Pacific
Finland 67
Forum (South Pacific) 23, 26
France 2, 34ff, 116, 119

GDR 82
Geneva International Peace Research Institute 141

143

INDEX

Germany (East) see GDR
Germany (West) see West
 Germany
Gorbatchev, M. 6
Great Britain 33
Greece 94, 96, 133
Green Party 87
Greenpeace 3, 35ff
 see also Rainbow Warrior

Hackett, J. 73
Hawke, R. 27, 34
Herman, J. 86
Hiroshima 1
Hungary 96

IAEA (International Atomic
 Energy Agency) 11, 17,
 30ff, 120, 123
India 2, 111, 116
Indonesia 39
Institute for Resource and
 Security Studies 141
Ireland 82
Israel 2, 84, 120

Jackson, G. 86
Japan 136

Kiribati 28
 see also South Pacific
Kirk, N. 25
Kossova 96
Krushchev, N. 83

Lange, D. 33
League of Nations 5
Libya 121
London Dumping Convention
 37
Low, D. 5

McNamara, R. 68
Macedonia 96
MAD (Mutually Assured
 Destruction) 72
Madriaga, J.S. 86
Malvinas 135
Medvedev, R. 87
Micronesia 27

Mills, J. 86
Miro, J. 86
Mountbatten, Lord 76
Mozambique 84
Muldoon, R. 26
Mururoa 3
MX Missiles 70

Nagasaki 1
Namibia 121
NATO 74
Neutron Bomb 70
New Zealand 2, 3, 25ff, 33,
 139
Nigeria 115
Niue 28
Nixon, R. 75ff
Noel Baker, N., Lord 80
Non-Proliferation Treaty 2,
 23, 37, 123, 138
Norway 83
 see also Europe
Nuclear Energy 16
Nuclear Free and Independent
 Pacific Movement 61

OAU 114, 119
Oceans 31ff
 see also South Pacific
OPANAL 13

Pakistan 2, 116
Palau 3
Palme, O. 68, 79
Papandreou, A. 87
Papua New Guinea 25, 40
Peace Movements 136
Poland 1, 82
 see also Rapacki
Policing 41
Priestley, J.B. 86
Pugwash Movement 87
Pym, F. 67

Rainbow Warrior 1, 3, 35
Rapacki, A. 1, 82, 116
Rarotonga (Treaty of) 5,
 23-66
 compared to Tlatelolco 33,
 47

INDEX

Reagan, R. 34
Rhodesia
see Zimbabwe
Robles, G. 4, 9ff, 124
Romania 82, 94
Russell, B. 85

SALT II 73
South Africa 2, 119, 122
South Korea 2
South Pacific 2, 23-45, 46-66
see also Cook Islands,
Kiribati, Tonga,
Vanuatu, Fiji, W.
Samoa, Papua New
Guinea
Soweto 122
SPREP (South Pacific Regional
Environmental Programme)
37
Sweden 83
see also Unden Plan
Switzerland 119

Taiwan 2
Tito, President 68
Tlatelolco (Treaty) 9-23
articles 13ff
compared to Rarotonga
33, 37ff, 111-13
Tonga 25
see also South Pacific
Transylvania 96
Turkey 94, 130

Tuvalu 26
see also South Pacific

U Thant 124
Unden Plan 2, 83
United Nations 4, 9, 14, 24,
115, 121, 126
panel on Nuclear-Free
Zones 42
special session on
disarmament 1978 80,
128
see also Non-Proliferation
Treaty
USA 3, 33, 67ff, 119
USSR 3, 33ff, 67ff, 111
military planning 78

Vanuatu 40, 46
Verification 17, 29
Vorster, J. 116

Warsaw Pact (WTO) 94ff
West Germany 67
Western Samoa 28
see also South Pacific

York, S. 86
Yugoslavia 68, 94

Zaire 121
Zimbabwe 84, 122
Zuckerman, Lord 72

145